Mass spectrometry and ion-molecule reactions

Cambridge Chemistry Texts

GENERAL EDITORS

E. A. V. Ebsworth, Ph.D.
Professor of Inorganic Chemistry
University of Edinburgh

P. J. Padley, Ph.D.
Lecturer in Physical Chemistry
University College of Swansea

K. Schofield, D.Sc.
Reader in Organic Chemistry
University of Exeter

Mass spectrometry and ion-molecule reactions

P. F. KNEWSTUBB
Assistant Director of Research
Department of Physical Chemistry
University of Cambridge

CAMBRIDGE
at the University Press, 1969

Published by the Syndics of the Cambridge University Press
Bentley House, 200 Euston Road, London N.W.1
American Branch: 32 East 57th Street, New York, N.Y.10022

© Cambridge University Press 1969

Library of Congress Catalogue Card Number: 69–16282
Standard Book Numbers:
521 07489 4 clothbound
521 09563 8 paperback

Printed in Great Britain
at the University Printing House, Cambridge
(Brooke Crutchley, University Printer)

Contents

Preface	*page*	vii
1	Reactions between ions and molecules	1
2	Preparation and reaction of an ion sample	21
3	Types of mass spectrometer	60
4	The use of mass spectrometry in problems of analysis	75
5	The interpretation and prediction of mass spectra	98
Note on Système International (SI) units		131
Bibliography and comprehensive references		132
Specific references		133
Index		135

To
K. F. SMITH

Preface

This book, being written as part of a series for third-year undergraduate or first-year research students, is deliberately of an introductory nature. The principal matters of interest are discussed at some length, while lesser off-shoots of the subject are given mention to indicate their possibility and their position in the field. For minor topics, or yet fuller treatment of any point, the reader is directed towards more specialised texts. It has also been my policy to include few rather than the maximum number of references to other work, in the belief that a great number of names, carrying little sense of the individuals referred to, serves only to confuse a reader at this stage.

The title of the book shows its reference to two topics which, at least until recently, had distinctly different approaches and ways of thought. The rapid development of each must surely lead to more and more common ground and, I believe, a real merging of the two fields.

Grateful thanks are offered to members of my family and to Miss P. Imrie for their assistance in the typing, to many others who helped in the development and preparation of the work, and particularly to Professor E. W. McDaniel for making available to me prior to its publication, information on and portions of the A.M.P.I.C. 'Review of Ion–molecule Reactions'. Thanks are also due for permission to use numerous figures from other sources, of which detailed acknowledgement is made in the appropriate captions. Last in this preface, though not so in my esteem, I mention with thanks the firm and thorough work of the editor, Dr P. J. Padley; any errors which remain cannot be such as he would correct, and must be laid at my door.

P. F. K.

1 Reactions between ions and molecules

1.1. A chemist's view of ion–molecule reactions. This book will attempt to show alongside the intimate relationship between mass-spectrometry and the reactions of ions with neutral molecules, the conditions under which each may be divorced from the other. In the earliest history of mass-spectrometry, in the work of Thomson, the ease with which ions could react with molecules was shown by the detection of ion currents which could not appear by direct ionisation of any species known to be present. Examples were the ions of mass-to-charge ratio $(m/e) = 3$ and 19, which could only be explained by postulating the ion-molecule reactions

$$H_2^+ + H_2 \rightarrow H_3^+ + H$$
and $$H_2O^+ + H_2O \rightarrow H_3O^+ + OH$$
or $$H_2^+ + H_2O \rightarrow H_3O^+ + H$$

or other similar reactions. These occurred so readily that the ion H_3O^+ could not be eliminated from any such experiment until considerable improvements of vacuum pumps and technique had been evolved. It is now of course quite possible with reasonable care to operate a mass spectrometer without 'interference' from ion–molecule reactions.

Looking at the subject from another angle, one may point to irrefutable evidence for the presence of ions in such régimes as electric discharges, flames, the upper atmosphere and in samples subjected to irradiation by high-energy particles. Generally in all these systems, the particle densities or times available are such as to afford ample opportunity for further collisions of the ions with neutral molecules, and interest may turn to the possibility of major observable effects from any reactions which may ensue. The debate rages perhaps most hotly at the moment in the field of radiation chemistry. Here a pure hydrocarbon may be irradiated, and numerous other hydrocarbons be discovered in the products. The divisive question is whether these arose mainly through reactions of neutral radical species, or of charged species. Whereas the former often have the advantage of higher concentrations, the latter may offset this by a higher rate of reaction. Very similar questions are relevant in the other systems mentioned. It is not surprising that mass-spectrometry is more and more being brought in to assist the investigations, but it did not create these problems.

It is one of the most notable features of reactions between ions and molecules that they are, with almost no exception, in the class of very fast reactions. As will be seen, this is a guiding factor in all attempts at measurements of reaction rates in this field. The reasons for its being so are also a topic for investigation and discussion.

The physical chemist may most readily appreciate the speed of a reaction by its rate constant, and some illustrative examples (for bimolecular reactions at 25 °C) are compared in table 1.1 to show the range of values covered.

TABLE 1.1. *Examples of bimolecular rate constants at 25 °C (approximate)*

Reaction	k in $\dfrac{\text{cm}^3}{\text{molecule sec}}$	k in $\dfrac{\text{litre}}{\text{g. mole. sec.}}$
Ion–molecule reactions:		
range of most values	$(40-0.4) \times 10^{-10}$	$(240-2.4) \times 10^{10}$
$H_2^+ + H_2 \rightarrow H_3^+ + H$	20×10^{-10}	1.2×10^{12}
$H_2O^+ + H_2O \rightarrow H_3O^+ + OH$	13×10^{-10}	7.6×10^{11}
$Ar^+ + H_2 \rightarrow ArH^+ + H$	3.5×10^{-10}	2.1×10^{11}
$N^+ + O_2 \rightarrow NO^+ + O$	5×10^{-10}	3×10^{11}
$O^+ + O_2 \rightarrow O_2^+ + O$	0.4×10^{-10}	2.4×10^{10}
Atom and radical reactions:		
$C_2H_5 + C_2H_5 \rightarrow C_4H_{10}$	2.7×10^{-11}	1.6×10^{10}
$O + NO_2 \rightarrow NO + O_2$	2.5×10^{-12}	1.5×10^9
$O + O_3 \rightarrow 2O_2$	2.5×10^{-14}	1.5×10^7
$H + H_2 \rightarrow H_2 + H$	7×10^{-17}	4×10^4
$CH_3 + C_2H_6 \rightarrow CH_4 + C_2H_5$	8×10^{-22}	5×10^{-1}
$Br + H_2 \rightarrow HBr + H$	8×10^{-24}	5×10^{-3}
Reactions of molecules:		
$NO + O_3 \rightarrow NO_2 + O_2$	2×10^{-14}	1.3×10^7
$HI + C_2H_5I \rightarrow C_2H_6 + I_2$	1.6×10^{-28}	1×10^{-7}
$2NO_2 \rightarrow 2NO + O_2$	1.6×10^{-31}	1×10^{-10}
Ions in aqueous solution:		
$e^-_{(aq)} + H^+ \rightarrow H$	4×10^{-11}	2.3×10^{10}
$CO_2 + OH^- \rightarrow HCO_3^-$	7×10^{-18}	4×10^3
$CH_3COOC_2H_5 + OH^- \rightarrow CH_3COO^- + C_2H_5OH$	1.6×10^{-22}	1×10^{-1}

1.1 A chemist's view of ion–molecule reactions

In the first place it may be noticed that the values of rate constant for ion–molecule reactions are practically encompassed in a range of two orders of magnitude. This observation is to be set alongside the postulate which has been made, that all exothermic ion-molecule reactions proceed with no energy of activation. Only in isolated cases among those so far investigated does this seem to be in doubt. It is also worth noting that, since most experiments are designed to measure very fast reaction rates, it could be partly for this reason that very few endothermic ion–molecule reactions, or any with an appreciable energy barrier, have so far been observed.

1.2. 'Cross-section' for a reaction. Even granting the absence of activation energy, the rapidity of ion–molecule reactions is such as to require immediate further comment. The explanation to be given in §1.4 will call for study of the reactions on an atomic scale, in which the energy of each collision is an important parameter of the event. The chemical rate constant, however, is a measure of the probability of reaction averaged over a Boltzmann distribution of energies corresponding to some temperature. For the detailed discussion, we will abandon this in favour of the physicist's concept of 'reaction cross-section'. This conveys the idea of the area presented by the target molecule to the ion, considered as an approaching projectile. If the latter is incident within the 'cross-section', reaction occurs.

From a more macroscopic point of view, a beam of ions carrying a current I may be considered to pass through a gas target of small thickness dx and density n molecules/cm^3. The small fraction reacting will be $-dI$, where

$$-\frac{dI}{I} = nq\,dx \tag{1.1}$$

where q is the appropriate cross-section. Hence in general for a target of any thickness x,

$$-\log I = nqx + \text{constant}$$

so that the beam current remaining at a distance x is

$$I = I_0 e^{-nqx}, \tag{1.2}$$

where $I = I_0$ at $x = 0$.

This is of course exactly analogous to Beer's Law for the absorption of light, and equally in both cases it is sometimes useful to consider the product

$$nq = Q,$$

where Q is the absorption coefficient (cm^{-1}) and may be tabulated for some standard density (e.g. 760 mmHg (torr) and 0 °C).

For ions, as for photons, the exponential law (1.2) will only be obeyed strictly for beams of particles of a single energy, and when only one interaction is possible. Granted these conditions, the reaction cross-section may be measured in three different ways:

(*a*) from the exponential decay of the original beam of ions along the path, assuming that non-reactive scattering and other effects can be discounted;

(*b*) from the corresponding appearance of products of the reaction along the path;

(*c*) (perhaps most often) from measurements of the ratio of product to reactant ion flux along the path.

The result of an ideal experiment of this type would be the 'microscopic reaction cross-section' (q) for some specific value of the incident ion energy. Repetition of the experiment at other values of ion energy would in general reveal a variation of q with this parameter (to be discussed later).

Such ideal experiments are not easy to perform, and it will be seen that the majority of results are obtained for incident ions with a range of energies, sometimes corresponding to a thermal (Boltzmann) distribution, sometimes to other distributions. The crude averaged result of such measurements is referred to as the 'experimental' or 'phenomenological' or 'macroscopic' cross-section, and is here called q_e. Such values require some considerable unravelling to give values of q (Light 1964). When referring to thermal energy distributions the values may be used more directly to give values of ('thermal') rate constant k. To a reasonable approximation one may use

$$k = q_e v \qquad (1.3)$$

where v is the mean relative velocity of ion and molecule, and k the rate constant in cm^3 molecule^{-1} s^{-1} units. More accurate derivations of k, with correct averaging, have been discussed but are often unjustified considering the other experimental errors.

The concept of 'collision diameter', which often appears in kinetic theory expressions, may also be defined for ion–molecule

1.2. 'Cross-section' for a reaction

reactions and related to q_e by

$$q_e = \pi\sigma^2 \times \frac{v}{\text{mean velocity of ion}} \qquad (1.4)$$

The collision diameter σ is in this case understood to refer only to collisions in which reaction occurs. The figure obtained in this way is of course an average over the energy distribution, just as much as is q_e.

1.3. Units of cross-section.
It should be noted that while q and q_e are often given in units of cm^2, the atomic units of πa_0^2 is also sometimes used, where a_0 = radius of the first Bohr orbit of hydrogen = 0.5292×10^{-8} cm. In the closely similar case of absorption of photons, the absorption cross-section is often given in Mb (1 megabarn = 10^{-18} cm^2).

The examples of rate constants for ion-molecule reactions which were given in table 1.1 are converted in table 1.2 to these other units, to show the magnitudes involved for these typical reactions. Absorption cross-sections for photoionisation are usually found to lie in the range 10–400 Mb.

TABLE 1.2. *Values of cross-section in various units for mean thermal velocities at* 300 °K (*Table* 1.1 *extended*)

Reaction	q_e (cm^2)	q_e (atomic units)	σ (cm)
Range of most values	$(1.6–200) \times 10^{-16}$	1.8–230	$(0.6–6.7) \times 10^{-8}$
$H_2^+ + H_2 \to H_3^+ + H$	76×10^{-16}	86	4.2×10^{-8}
$H_2O^+ + H_2O \to H_3O^+ + OH$	150×10^{-16}	170	5.8×10^{-8}
$Ar^+ + H_2 \to ArH^+ + H$	19×10^{-16}	22	2.1×10^{-8}
$N^+ + O_2 \to NO^+ + O$	60×10^{-16}	68	3.7×10^{-8}
$O^+ + O_2 \to O_2^+ + O$	5×10^{-16}	5.7	1.1×10^{-8}

1.4. Measurement of cross-sections.
Besides the cross-sections for ion-molecule reactions and for photoionisation, the field of mass-spectrometry deals also with cross-sections for ionisation of a molecule by impact of an electron, this being the method most frequently used for the generation of ions for mass analysis. The impact of heavy particles, charged or neutral, of high energy (> 1 keV), is a topic falling only on the fringes of a study of ion-molecule reactions in its presently accepted sense.

As will be seen, the mass-spectrometer itself often provides one of the best ways of measuring *relative* cross-sections for similar reactions, or for variation of some parameters such as electron energy, but it is not ideally suited to the absolute measurement of cross-section.

The method most used for absolute measurements is well illustrated by the apparatus of Tate and Smith, in which the ionisation of various gases by a beam of electrons of controlled energy was studied (see fig. 1.1).

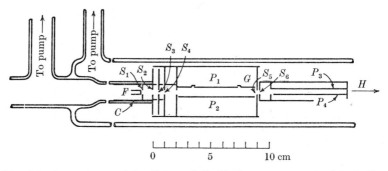

Fig. 1.1. Apparatus used by Tate and Smith for measurement of ionisation potentials and apparent ionisation cross-sections. H signifies an axial magnetic field; other explanations in the text. [From E. W. McDaniel (1964) p. 177. *Collision Processes in Ionised Gases*. John Wiley and Sons, Inc. Originally from Tate & Smith (1932) *Phys. Rev.* **39**, 270.]

The apparatus is evacuated to 'high vacuum' (10^{-5} torr or less). Electrons are emitted from the filament F, and given a controlled acceleration to a known energy by application of suitable potentials to the electrodes. The slits S_1–S_4, besides providing these potentials, intercept electrons not travelling precisely in the desired direction. This collimation of the beam is in fact greatly assisted by the axial magnetic field which is applied, since electrons with any component of velocity perpendicular to the field undergo a type of spiral motion which prevents any great divergence of the beam from its axis. Thus the electrons are unable to strike the plates P_1, P_2 but are collected on P_3 and P_4. A weak transverse electric field between P_1 and P_2 allows the collection on P_1 of any positive ions formed in a known length of the electron beam (a length equal to the length of P_1, since guard plates G are maintained at the

1.4. Measurement of cross-sections

same potential as P_1). The electron beam current and the gas pressure are also measured. Since the attenuation of the electron beam is quite low at the gas pressures used, the calculation of cross-section follows from the simple formula (1.1).

Various improvements of this method have been described, notably by Bleakney and Lozier (q.v. Kiser; McDaniel) but the principle remains the same, and is used also for the study of cross-sections for ionisation of gases by photons, (and also by energetic heavy particles).

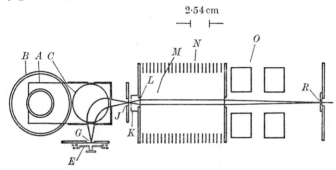

Fig. 1.2. An apparatus for measuring the cross-sections of ion–molecule reactions. A, B, and C are the yoke, coil and pole pieces of the source mass spectrometer; E its ion source; G its object slit; and J, its exit slit. The reaction chamber K is shielded by a grid L from the region M which is surrounded by a series of rings N that provide the boundary condition for a uniform accelerating field. The electrostatic lens O is used to focus the ion beam on the object slit R of the analysis mass spectrometer. [After C.F. Giese & W. B. Maier (1963) *J. chem. Phys.* **39**, 739.]

For the study of ion–molecule reaction cross-sections, interest is greatest in the experiments where very low ion energies are used. This raises difficulties peculiar to this type of investigation, as the velocities of the reactant ions are comparable with those of the product ions, and mutual scattering will occur. The collection and measurement of the appropriate ion currents is most difficult to do in a reliable quantitative manner, but recently experiments have been performed successfully down to energies of 1 or 2 eV. A simplified diagram of the apparatus used in these experiments is shown in fig. 1.2 and shows first the selection of the desired ion at an appropriate energy by a small 'source mass spectrometer' (cf. fig. 3.2). These ions then enter a small collision chamber, of known length and containing a known pressure of gas. The ions leaving the chamber are collected and focused into an 'analysis mass spectrometer' by carefully designed electric fields. In some experiments of this type (Maier), endothermic ion-molecule reactions have been observed, with cross-sections rising abruptly from an apparent energy threshold of several tenths of an electron volt in the bombarding ion.

1.5. The Langevin theory of collisions of a charged particle.

The reasons for the large cross-sections found for ion–molecule (or ion–ion) collisions were first given detailed quantitative treatment by Langevin, (see McDaniel) who was interested in the mobility of ions in gaseous media. His theory is now seen as a fundamental contribution to the understanding of ion-molecule reactions, and some consideration of it will now be given, though with the minimum of mathematical detail.

The theory rests on consideration of the long-range attractive forces produced if the approaching ion is able to induce appreciable polarisation of the target molecule (the more so, if a permanent dipole is also present). If these forces are sufficiently strong, and the velocity of the ion relative to the 'target' molecule not too great, the attraction will cause the closest approach of the two particles r_a to be considerably smaller than the 'impact parameter' b, which would be the closest approach of the particles in the absence of any interaction between them (see fig. 1.3a).

The following outline of the theory follows that to be found in McDaniel's book. The ion at a distance r from the molecule induces a dipole of $\alpha e/r^2$ where α is its (mean) polarisability. The potential energy of the ion in the field of the dipole is then

$$V = -\frac{\alpha e^2}{2r^4} \qquad (1.5)$$

e being the electronic charge. At close range there must be a repulsive (hard core) potential, giving as a possible complete potential for the interaction

$$V(r) = \frac{a}{r^{12}} - \frac{\alpha e^2}{2r^4} \qquad (1.6)$$

It is now convenient to add to this a fictitious potential term which expresses in the same units the centrifugal effect which is important in the collision. Considering that the angular momentum of the

1.5. Langevin theory of collisions of a charged particle

system j will be conserved in the collision, one derives a centrifugal potential

$$V_c(r) = \frac{j^2}{2Mr^2} \quad (1.7)$$

(M being the reduced mass of the system). By adding $V_c(r)$ to (1.6), the centrifugal effects are so represented that one may consider

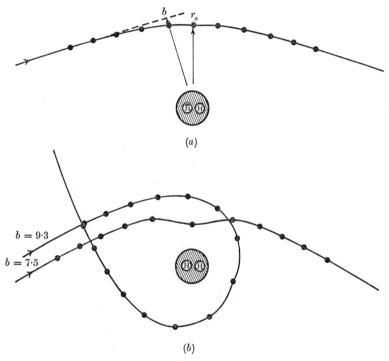

Fig. 1.3. (a) Example of a 'distant' collision. $b = 10$ a.u. (b) Examples of 'close' collisions. $b = 9 \cdot 3$ and $7 \cdot 5$ a.u. (Critical $b_0 = 9 \cdot 4$ a.u., time intervals 5×10^{-14} s), calculated for the interaction $H_2^+ + H_2$.

the particles to interact in a one-dimensional system under the combined potential; this is readily shown on a diagram. The curves of fig. 1.4 have been calculated with parameters appropriate to the collision of H_2^+ with H_2, to provide a definite example.

The lowest curve is for the interaction of ion and molecule with no effects of angular momentum. To this the term $V_c(r)$ was then added, assuming a collision energy of $3/2\,kT$ (for 300 °K) and the stated impact parameters. The same curves would of course arise for any other combinations of energy and impact parameter which gave the same values of angular momenta. The collision energy $3/2\,\mathrm{k}T$ is also marked, and shows that the collision with $b = 10$ a.u. (atomic units) is of a different character from that with $b = 9\cdot3$ a.u.

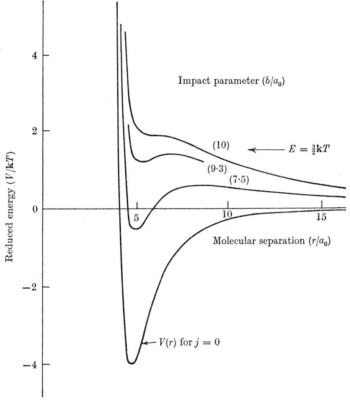

Fig. 1.4. Effective potential curves for 'high j case' ($b = 10$), 'low j case' ($b = 7\cdot5$) and intermediate ($b = 9\cdot3$), drawn for $H_2^+ + H_2$. Energy relative to $\mathrm{k}T$ at 25 °C, atomic units (a_0) used.

(or 7·5 a.u.). Whereas in the first case the nuclei approach no closer than 8·6 a.u., in the others the minimum distance of approach is respectively 4·73 a.u. and 4·4 a.u.

The detailed mathematical development of the theory enables a calculation of a critical value of impact parameter b_0 (for a given

1.5. Langevin theory of collisions of a charged particle

collision energy or relative velocity of approach), below which the distance of closest approach r_a changes discontinuously to a much smaller value. For the numerical example discussed above, $b_0 = 9\cdot4$ a.u., and if this condition were exactly achieved, the particles would enter a metastable orbit around each other. For values of b close to b_0, the trajectory of the ion around the molecule will take on a spiral or 'orbiting' nature, and the duration of the interaction is greatly increased. The trajectories have been calculated for the example taken, and are shown in fig. 1.3; the 'orbiting' effect is evident for $b = 9.3$ a.u. Collisions may be divided into 'close' or 'distant' by the relation of b to the critical value b_0.

In applying Langevin's theory to ion–molecule reactions, the key assumption was made that the occurrence of a spiral-type trajectory or a 'close' collision leads to reaction, while a simple 'distant' deflection of the ion path does not. The critical value of impact parameter b_0 is then calculated and equated to the collision radius of the system. The results appeared encouraging in earlier work, but with the refinement and extension of experiments (and increasing ambition of experimenters), further elaborations of the theory are now being pursued (A.M.P.I.C. Review).

The mathematical treatment, which is not reproduced here, gives for the critical impact parameter

$$b_0 = \left\{\frac{4\alpha e^2}{Mv_0^2}\right\}^{\frac{1}{4}}$$

$$= \left\{\frac{2\alpha e^2}{E}\right\}^{\frac{1}{4}} \quad (1.8)$$

where v_0 is the relative velocity of approach at large separations and E the corresponding energy. Hence the cross-section for collisions of 'close' type is found to be

$$q = \frac{2\pi}{v_0}\left\{\frac{\alpha e^2}{M}\right\}^{\frac{1}{2}} \quad (1.9)$$

The predictions of the theory as regards time of collision are illustrated in fig. 1.5, which is a further approximate calculation for the case of $H_2^+ + H_2$, for the collision parameters stated. It might be expected that one factor besides closeness of approach which determines the 'success' of a potentially reactive collision is the *time* during which the molecules are within some critical distance of each other. This time is shown as a function of impact parameter.

The indeterminately large times indicated in the upper two curves arise from the possibility of metastable orbits. However, since the critical impact parameter of 9.4 a.u. would give a metastable

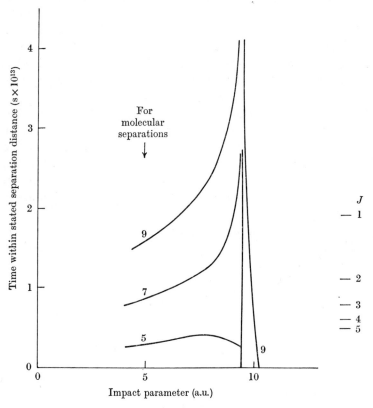

Fig. 1·5. Predicted time of collision versus impact parameter. Note the critical parameter at 9·4 a.u.

orbit of radius 6·6 a.u., this effect is not seen in the lowest curve. It is also apparent from the figure that some 'distant' collisions ($b > 9 \cdot 4$ a.u.) result in $r_a \leqslant 9$ a.u., but that all such events are sharply excluded from the curves for $r_a \leqslant 7$ a.u. or 5 a.u. The scale at the right of fig. 1·5 (where J refers to rotational quantum number), shows that the collision times are all of the order of rotational

1.5. Langevin theory of collisions of a charged particle

periods of these quantum states for H_2, calculated classically from the known momentum. As is pointed out in the next section, however, the pursuit of great detail in this manner is very difficult or even impossible.

1.6. Extensions to the Langevin theory. The key assumption mentioned above, that 'close' collisions will lead to reaction while distant collisions will not, may seem more than a little arbitrary. It can be rationalised to some extent by supposing that for any ion-molecule system there exists a critical distance of approach r_c which is the true criterion of success of the reaction. It will then be true for some range of energy that r_c lies in the gap between values of r_a (the closest approach) for 'close' collisions and those for 'distant' collisions. For that range of energy the original assumption is then justified. The implicit assumption that r_c is not dependent on the energy in the collision is in line with the general assumption that ion–molecule reactions proceed without activation energy.

The foregoing indicates that the simple application of the 'slow ion' theory is somewhat restricted, and attempts have been made to extend the line of argument to collisions at somewhat higher energies, at which reactions do occur, although the condition $b \leqslant b_0$ cannot be maintained since b_0 becomes so small. The mathematical treatment of the orbits can be used quite generally to give, even for collisions with slight deflexion of the ion path, the value of b for which $r_a =$ some constant value r_c. The resulting cross-section is of the form

$$q = \pi r_c \left\{ 1 - \frac{\text{Potential energy at } r_c}{\text{Initial relative kinetic energy}} \right\} \quad (1.10)$$

and thus tends to the value πr_c^2 at high energies. The argument can equally well be applied to repulsive interactions, and has had some degree of success.

A slightly different idea has been adopted (Lorquet and Hamill) in considering that the cross-section of the 'target' molecule predicted by Langevin theory (q_L) has a 'bull's eye' of constant cross-section (q_K) which represents the hard-core of the molecule. Since $q_L \propto E^{-\frac{1}{2}}$, the target 'shrinks' as the energy is raised, until at a transition energy E_t only the constant cross-section q_K is left. Further refinements are added for the differing probabilities of reaction in an orbiting collision and in a hard-core collision, and the resulting formulae fit quite well the data to which they are applied. However, it may be possible to apply other explanations to the same data (Chan & Sedgwick; Dugan & Magee).

A further type of attempt (Böhme, Hasted & Ong) to explain experimental data on cross-sections for reactions involving only atoms and diatomic molecules considers the 'time of collision' (t_c), somewhat empirically defined, in comparison with a 'transition time' (t_t) equal to $h/\Delta E$, where ΔE is the change of internal energy in the reaction. For charge transfer between neutral atoms, in which ΔE has essentially a unique value, it appears reasonable to say that the cross-section reaches a maximum for $t_c = t_t$, and the same idea is carried over to the more complicated case, in which ΔE may have many values considering all possible combinations of states of reactants and products. Now since in ion-molecule reactions at low energy the values of t_c may be quite long, particularly for orbits which 'spiral' considerably, the cross-section for reaction is expected to be large only when ΔE is very small. Hence a search is made for 'near resonance' conditions in the states of reactants and products (which must all be known with good accuracy—hence the restriction on species) and a total rate of reaction is estimated. The results appear to fall reasonably in line with the experimental data in those cases where the method can be applied.

In spite of the successes which have been reported, the theories at present do not carry confidence that they will prove of general application. A definite point of doubt in the Langevin theory lies in the use of the inverse fourth-power potential down to distances on the atomic scale, and the neglect of any anisotropy of the polarisability. There is a notable omission in all but the near-resonance theory of any consideration of energy transfer to (or from) the internal motions of the colliding molecules. One would expect that a molecule experiencing an orbiting collision, for example, would show considerable changes of its rotational energy. Finally it must be acknowledged that attempts to apply the Langevin theory with good accuracy are likely to fall foul of the uncertainty principle, and be unlikely to produce trustworthy predictions in detail. Since matter has an underlying wave nature, it must be recognized that detail of motion finer than the De Broglie wavelength of a particle ($\lambda = h$/linear momentum) cannot truly be resolved. In the

1.6. Extensions to the Langevin theory

case of the numerical example discussed above, this quantity is of the order of 3 a.u. Another view of the same difficulty may be seen in the rotation of a pair of particles round each other, where the angular momentum may be exactly known (and will be quantised) but where there will be concomitant uncertainties in momentum and position (and hence in moment of inertia) which make the calculation of exact orbits of doubtful validity.

1.7. The phase-space theory. The 'phase-space' theory of ion–molecule reactions was first put forward by J. C. Light (1965, 1968), and is currently receiving detailed evaluation. It is based on the principles of statistical thermodynamics, and concerns itself with the most probable distribution of the products of a collision among the available energy states. These states include the various electronic, vibrational and rotational energy levels as well as translational energy, and include the possibility that the 'products' are chemically the same as the 'reactants', i.e. that no chemical reaction has occurred.

The behaviour of the system may be specified in simplest (though not most useful) terms by coordinates of momentum and position, which number six for each nucleus. Thus in a collision of an atomic ion with a diatomic molecule the products emerge in a state which is completely specified by lying within a small 18-dimensional element of the so-called 'phase space'. If it is then supposed that the total energy and the total angular momentum vector are given, the necessary conservation of these throughout the collision imposes four restrictions on the elements of phase space which may finally be occupied. The vibrational state of the diatomic molecule in both reactants and products may also be treated as known if the calculations are repeated for all likely combinations of these. The elements of the phase space which are still 'allowed' under these restrictions are then assumed to be occupied randomly when a large number of collisions is studied. The relative phase space volume belonging to each possible set of products may then be determined, and is taken to predict the ratios in which the products will be found.

The total energy of the system will include any vibrational or rotational energy of the initial particles as well as the relative translational energy, and the impact parameter is involved in the calculation of initial angular momentum. Other than this, it is

the essential assumption of the theory that all information about
the initial states of the colliding particles (e.g. direction and magnitude of velocity) is lost when the collision complex has formed.
The system must 'have no memory' of its initial state, if a random
distribution over the available elements of phase-space is to occur.
This is the salient difference between the phase-space theory and
the classical (Langevin) type of theory. However, the ideas of the
latter are brought in by supposing that only the 'close' collisions
satisfy this 'strong coupling' condition. The cross section πb_0^2 then
acts as a normalising factor, i.e. it is 'shared out' amongst the
possibilities as suggested by the phase-space theory.

The actual computations required for exploration of even the
simplest cases, i.e. reactions involving an atom and diatomic
molecule, are such as to demand the use of machine techniques.
Such calculations have been done for a few cases, in some of which a
comparison with experimental data is possible. One example is
displayed in fig. 1.6 for the reaction

$$N^+ + O_2 \to \text{products}$$

where the curve A agrees tolerably well with the experimental
curve EXP which was derived from measurements of NO^+ produced
in the reaction. The calculations suggest the proportions in which
other possible products should appear. They also predict that
endothermic processes such as dissociation into three atoms (D)
and the 'reverse' reaction (B) (meaning no observable reaction)
should occur with energies above a certain threshold value.

Perhaps the most fair general statement is that the phase-space
and Langevin types of theory are opposite extremes, and that the
true situation may be variously a mixture of both. The 'strong
coupling' condition of the phase-space theory may apply well for
example, to rotational degrees of freedom but not to vibrational
degrees during an 'orbiting' collision, while the reverse may be true
for a 'head-on' collision. The whole question of the theory of these
collisions is a very active subject at present, and further developments are to be expected.

1.8. Simplification and averaging in cross-sections. The
concept of cross-section is applied not only to ion–molecule reactions, but also to the production of ionisation and excitation in a

1.8. Simplification and averaging in cross-sections

collision, and elastic scattering. The 'total collision cross-section' is used for the sum of all these effects.

The discussion in the previous sections may be taken to illustrate some of the simplifications involved in defining a cross-section. If information about the collision were available in full detail, it would take the form of 'probability of the desired event' (be it

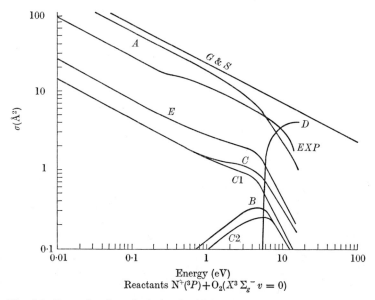

Fig. 1·6. Example of a calculation based on the phase space theory. Products NO^+, O^+, O_2^+ from the reaction $N^+ + O_2$. The experimental results shown are those of Stebbings et al. Extrapolation is made to the low-energy experimental data of Fehsenfeld et al. Products are as follows: A, $O+NO^+$; EXP, $O+NO^+$; D, dissociation, $O^+(^4S)+NO$ $(X^2\pi_r)$; B, $N^+(^3P)+O_2(X^3\Sigma_g^-)$; C, $N+O_2(X^2\pi_g)$; $C1$, $N(^4S)+O_2^+$; $C2$, $N(^2D)+O_2^+$; G & S, refers to the Langevin cross-section (πb_0^2). [From F. A. Wolf (1966). J. chem. Phys. **44**, 1619].

reaction to certain specified products, production of ionisation, or of excitation, etc.) as a function of some initial parameter or parameters.

For most cases of ion-molecule reactions, one would wish for a plot of 'probability of reaction' against impact parameter b, for constant initial energy or velocity (and this repeated at other values of energy). This could take the form of the notional sketch of fig. 1.7.

In assigning a definite cross-section (q), or collision diameter (σ) to the event, one is constructing a rectangle on such a plot, of height unity and area equal to the area under the sketched curve. In the sample of fig. 1·7 the critical parameter of the theory (b_0) is a fair measure of the collision radius, but one can easily imagine less favourable cases.

If it is allowable thus to discuss the detail of individual events, one may note that it is not even necessarily the case that the plot

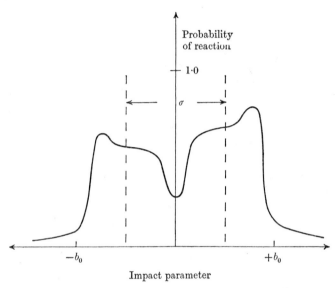

Fig. 1.7. Notional sketch of probability of reaction versus impact parameter.

of probability is symmetric about the centre of mass of the target particle. The interaction of an orbiting ion with a rotating molecule can depend on the sense of rotation. As soon as one allows the discussion of particles of other than spherical symmetry, it becomes possible that the cross-section as defined above becomes a function of all the possible relative attitudes of the particles during the collision, and effects observed must be considered to be an average over a random occurrence of such possibilities.

It should from this be realised that the cross-section concept, while most useful, does represent considerable averaging of information, the detail of which is not available. No experiment has

1.8. Simplification and averaging in cross-sections

yet been devised in which the impact parameter is a controllable variable and there seems to be little prospect of such. The question of interaction between molecular rotation and orbiting has been examined in the case of molecules having a permanent dipole. A theoretical study (Dugan & Magee) indeed suggests a strong possibility that at low ion energies the two motions tend to become 'locked together', and this appears to be supported by features of experimental work with molecules such as CH_3CN (Moran & Hamill).

In conclusion the reader is brought back to the first topic of this chapter, the chemical rate constant for a reaction. A quantity having the units of a rate constant is given by the product of cross-section and relative velocity

$$k' = qv$$

for any collision event. To a chemist, however, rate constants are best applied to describe the behaviour of systems of large numbers of particles when the energy distribution in most, if not all, degrees of freedom is approximately of a Boltzmann form (i.e. a situation to some degree in equilibrium at some temperature). Supposing that the variation of q with energy were known, a complicated procedure might be derived to produce values of chemical rate constant for some given Boltzmann distribution of energy.

In fact the situation is in many cases not quite so complicated. Examination of (1.9) shows that the Langevin 'slow ion' theory predicts a value of qv which is independent of energy, and hence would give immediately a correct and meaningful value of rate constant. In practice this is not exactly the case, and the procedure adopted is to plot the product qv against energy and extrapolate to very low (thermal) energies to obtain a value which is declared as the 'thermal rate constant'. In experiments in which a value of cross-section is obtained which is already an average over a range of energies, q_e, the product of this with the mean velocity in the measurement may be treated similarly, as was indicated in equation (1.3).

The true chemical definition of a rate constant, in terms of the rate of disappearance of a reactant (primary ion) or of appearance of product (secondary ion) can be brought directly into the treatment of experimental data in certain types of experiment. Thus for a reaction

$$A^+ + xB \rightarrow C^+ + D$$

the rate constant is to be expressed by

$$k = -\frac{1}{[B]^x}\frac{\mathrm{d}\ln[A^+]}{\mathrm{d}t} = \frac{1}{[B]^x}\frac{\mathrm{d}\ln[C^+]}{\mathrm{d}t}$$

where quantities in square brackets imply number densities. It is possible to use this approach when the ions exist for a measurable time at a constant energy, or in a constant energy distribution, together with neutral molecules in thermal equilibrium at some known temperature. A direct determination of a 'thermal rate constant' requires also that the ions be (at least, approximately) in thermal equilibrium with the neutral molecules, which generally implies the absence of all but weak electric fields from the reaction region. Examples of such experiments may be found in §§ 2.3 and 2.4 pp. 48 and 50.

The question of temperature-dependence of rate constants, so much in evidence in other fields of chemistry, has not yet received much attention for ion–molecule reactions. If these reactions do indeed proceed without activation, then no marked variation of values would be expected.

2 Preparation and reaction of an ion sample

2.1. Introduction. This chapter concerns itself with two different aspects of the generation of ions from neutral molecules. The first is the viewpoint of the analyst, who will seek to generate a population of ions which, if not in exact correspondence with the neutral population may, if accurate measurement is achieved, lead to reliable inference of the original constitution of neutral species. Since the sample dealt with may be in solid, liquid or gaseous form, there are some considerable problems in this topic, and the possible occurrence of ion-molecule reactions has sometimes been listed amongst them.

On the other hand, there is a fast growing interest at this time in the further discovery and more exact study of ion–molecule reactions. The fields of interest which have prompted such studies were mentioned in §1.1 and are considered further in §2.3. For detailed research in this direction (see §2.4) one may use any method, any gas mixture which will give the required amount of reacting ion (and preferably no others) in the right conditions for study of its further reaction.

In all discussion in this chapter, the mass analysis of the ions produced is assumed without further comment; discussion of this is to be found in chapter 3.

2.2. Ion sources used in mass spectrometric analysis. *The electron impact source.* This seems to be the most versatile and widely used of all the types of ion source, and is restricted in its application chiefly by the necessity of presenting the sample as a vapour or gas at a pressure of 10^{-5} to 10^{-6} torr.

Fig. 2.1 shows a typical arrangement of this type of source. Electrons are emitted from a hot filament (commonly tungsten at *ca.* 2000 °C) and are accelerated and collimated by slits into a narrow beam crossing the axis of the source. The collimation and guidance of the beam is usually assisted by application of a magnetic field of about 100 gauss; low energy electrons will follow the lines of force of such a field in tight spirals. The electrons passing out of the source are captured by a 'trap' electrode. The 'nominal energy' of the electrons crossing the central portion of the source is set by the mean potential of the filament relative to the source. This may have any desired value, but is usually in the range 10–100 eV, while the

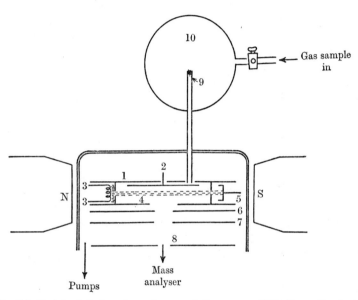

Fig. 2.1. Sketch of electron impact source. 1, Source case (0 V); 2, repeller electrode (+2 V); 3, filament (−70 V, 20 μA emission); 4, electron beam (70 eV energy); 5, electron trap; 6, draw-out electrode (−10 V); 7, acceleration electrode (−2,000 V); 8 analyser entrance slit (−2,000 V); 9, 'molecular' leak; 10, low-pressure reservoir (2×10^{-2} torr). (Potentials shown are typical, relative to source case.)

actual energies of the electrons are spread over a range of 1–2 eV, due to the high temperature of the filament.

Samples in gaseous form are admitted to the reservoir at a measured pressure in the range 10^{-2}–10^{-1} torr, and leak slowly into the ionising region of the source to give pressures of a few microtorr (up to a maximum between 10^{-5}–10^{-4} torr). Ions are then formed in proportion to the product of the relevant partial pressure and ionisation cross-section. Provided that the sample gases (if a mixture) enter and leave the source under conditions of molecular flow (i.e.

2.2. Ion sources used in mass spectrometric analysis

through channels which are always of considerably smaller dimension than the local mean free path) the relative partial pressures in the source will correctly reflect those in the reservoir. This condition governs some features of the design of the source, and the choice of reservoir pressure.

Although the ionising region should ideally be free of electric fields so that the electron beam and the resulting ion beam have well defined energies, it is often in practice advantageous to introduce the repeller electrode (fig. 2.1). This, together with field penetration through the ion exit slit, urges the positive ions in the right direction, while the electrons, due to their much higher velocity, are effectively constrained rather to follow the direction of the magnetic field and reach the electron 'trap', with relatively slight effect of the repeller field. This additional field thus gives a great gain in sensitivity and in uniformity of collection of variously formed ions at the cost of a small loss of mass separation—an acceptable exchange. It will also be seen to reduce 'interference' by ion-molecule reactions at the higher source pressures (see p. 45).

The ionisation of a molecule RX to give the 'parent ion' RX^+ occurs by the process

$$e^-(\text{fast}) + RX \rightarrow RX^+ + 2e^-(\text{slow})$$

but very often there are in addition one or more fragmentation reactions

$$e^-(\text{fast}) + RX \rightarrow R^+ + X + 2e^-(\text{slow})$$

The cross-sections (and hence the ion currents) vary with electron energy in a manner broadly the same for all molecules, as shown in fig. 2.2.

Since the energy spread of the electron beam about its nominal value will rarely exceed 2 eV (most of this being due to the high temperature of the filament from which it is emitted), the measured cross-section is very close to the microscopic cross-section q, except at the lowest energies. The curve for 'parent' ion RX^+ rises from a threshold which is (approximately) the ionisation potential (I.P.) of the molecule, while for each fragment ion, the threshold of the curve is related to the appearance potential (A.P.) of that ion from RX. The values of A.P. always lie at or above the relevant I.P., but it is by no means the case that the parent ion is the most abundant at all electron energies. Usually the curves show little slope in the vicinity of 70 eV electron energy, and this is often taken as a suitable value for analytical work, giving highest sensitivity and stability. For a molecule giving several fragments, the largest peak at the chosen electron energy is generally denoted 'base peak' and the others are then related to it on a percentage basis. Typical values of maximum cross-section lie in the range

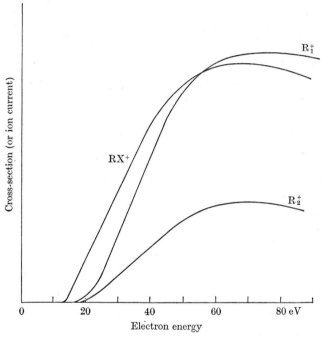

Fig. 2.2. Variations of cross-section (or ion current) with electron energy–parent and fragment ions (sometimes called electron-ionisation efficiency curves).

$(1-4) \times 10^{-16}$ cm². Hence one may see that the absorption of the electron beam in a 1 cm ionising region, even at the highest pressures used (10^{-4} torr), will be only 0·02–0·08 per cent.

Elaborations of the electron impact method. The accurate determination of the thresholds of the curves of fig. 2.2 should give good values of ionisation potential of RX (but see §5.2, and for appearance potentials see the note at the end of §5.7) which are often of interest. In the source as described, the threshold of the ionisation is marred by a curved foot due in part to the thermal spread of elec-

2.2. Ion sources used in mass spectrometric analysis

tron energies. A simple extrapolation of the upper part of the plot, ignoring the foot, can give a useful estimate of the threshold. Sometimes a plot of log (ion current) *vs* nominal electron energy is used, to 'straighten out' the foot. Because of contact potentials and other effects, the nominal electron energy is never the true value, and it is always necessary to plot also data from a gas of known I.P. and determine a potential difference.

The distortion at the threshold has been reduced or eliminated by two different methods. In some elaborate sources, the electron beam is passed through a small energy selector (compare §3.3) which acts like an optical monochromator. In others, the so-called retarding potential difference (RPD) method is used. Extra grids are added to control the electron beam, and one of these, with a retarding potential, allows only electrons with a thermal contribution to the total energy greater than a certain value to pass. A reduction of this potential by as little as 0·05 V can cause a measurable increase of ion current, and this change may be considered due to the additional 'slice' of electrons which can now pass, the energies of which lie within a narrow range of 0·05 eV. The change of ion current is thus associated with electrons which are almost 'monochromatic'. By either of these means the ionisation efficiency (cross-section) curve for parent ions can, for atomic species at least, be made to show a sharp threshold from which good values of the ionisation potential can be derived, again by comparison with a calibration gas.

The convenient features of the electron impact source have led to attempts to extend its use to other than gaseous or volatile substances, and a great deal can be done in this direction by having the whole reservoir and sample handling system heated to as much as 300 °C. When this fails, the material to be analysed is sometimes coated on a broad filament which acts both as the repeller electrode and as a heated element from which material may evaporate directly into the ionising region. In a more closely controlled version of this, the substance is placed in a heated effusion (Knudsen) cell, the small orifice of which is close to the electron beam. The ion currents which are then observed should be a measure of the partial pressures of vapours present *inside* the effusion cell, under conditions of thermal equilibrium. This topic is discussed further on p. 95 and illustrated in fig. 4.4.

The photon impact source. An ion source in which molecules are bombarded with high energy photons bears many resemblances to an electron-impact source. As is shown in fig. 2.3. (which may be compared with fig. 2.1), the ion source is attached to the exit slit of a vacuum monochromator which is usually of a type employing a diffraction grating ruled on a concave mirror of radius $\frac{1}{2}$–1 m.

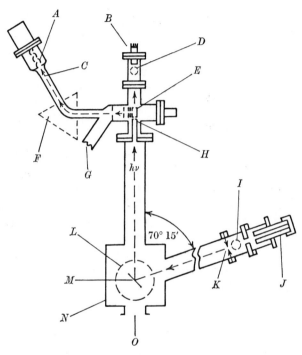

Fig. 2.3. Complete mass spectrometer with photoionisation source. *A*, Electron multiplier ion detector; *B*, photomultiplier radiation monitor; *C*, ion beam; *D*, *G*, *I*, *O*, to pumping systems; *E*, ion source; *F*, permanent magnet; *H*, monochromator exit slit; *J*, light source; *K*, monochromator entrance slit; *L*, grating turntable; *M*, grating; *N*, Seya-type vacuum monochromator. [From G. L. Weissler et al. (1959) *J. Opt. Soc. Am.* **49**, 338.]

The source of radiation will normally be an electric discharge dissipating at least 50 W through hydrogen or one of the rare gases. Since a considerable pressure of gas is needed for the best light output from the discharge, this is in some cases separated from the monochromator by a window of LiF. Even this, the best material available, severely limits the applications, as it is not transparent

2.2. Ion sources used in mass spectrometric analysis

to wavelengths below 1000 Å (equivalent to about 11 eV). The extension to lower wavelengths requires coupling via a small hole with strong pumping to maintain the necessary degree of vacuum in the monochromator. All this tends to generate a fairly bulky and expensive piece of apparatus.

Having indicated some difficulties in the use of this ion source, its advantages must be pointed out. The chief of these is a very

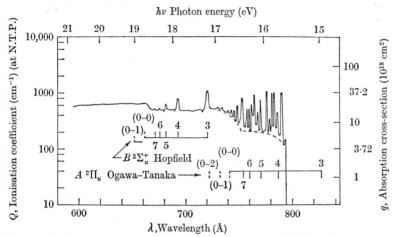

Fig. 2.4. Variation of cross-section for photoionisation of N_2. *Note*: In this figure energy increases toward the left. [After G. R. Cook & P. H. Metzger (1964) *J. Chem. Phys.* **41**, 329.]

close determination of ionisation potential (often to ± 0.02 eV) and observation of other details of the ionisation process. It is also true that there is little fragmentation because wavelengths quite close to the threshold are generally used; this might have advantages for some analytical purposes. It is important to note that, in contrast to the cross-section for electron bombardment ionisation (see fig. 2.2), the cross-section for photoionisation should theoretically rise abruptly from threshold to a fairly constant value at shorter wavelengths. This type of behaviour is generally found to be the case, with sometimes some additional features due to autoionisation (see fig. 2.4, for example). The value of cross-section in the working region generally lies in the range 10–60 Mb, with occasional values up to 400 Mb at specific wavelengths due to the autoionisation effects. These figures, together with the relatively low photon flux obtained from the monochromator, lead to ion currents considerably lower than those from electron impact sources.

The field ionisation source. This method of generation of ions is still fairly new (Gomer) and in the process of more extensive testing, but shows good promise for analytical work. The mechanical arrangement is very simple, involving essentially a finely pointed conductor, which may be a 'whisker' formed or etched by electrochemical means, or as in more recent work a very fine metal wire or razor-edged plate (Beckey *et al.*). This carries a high electric

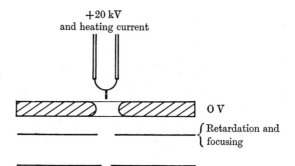

Fig. 2.5. Sketch of field ionisation source using a single 'whisker'. The sharp tip, mounted on a filament (for heating as required) is much finer than can be drawn.

potential, and is suitably surrounded by earthed surfaces (fig. 2.5). The arrangement produces a very high local electric field at the smallest radius of curvature (the whisker tip, wire or edge) which must in fact reach values of the order of 2×10^8 V/cm (2 V/Å). In the case of even the finest wires and edges available, the observed success of the method must be dependent on the occurrence of local small radii of curvature due to roughness of the surface or the growth of whiskers by some conditioning process, since the general radius is too large.

Having implied that field ionisation is likely to occur when local electric fields of about 2 V/Å are produced at a conducting surface, the manner of its occurrence must be suggested, even if in somewhat simplified terms. As a model situation, one may consider a hydrogen

2.2. Ion sources used in mass spectrometric analysis

atom entering the electric field, and commence with the idea of Bohr's theory of electron orbits.
As indicated in fig. 2.6(a), the potential energy of the electron in the various allowed circular orbits is constant at certain levels for the normal case of an atom remote from other influences. On

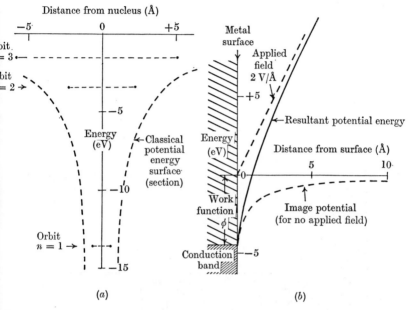

Fig. 2.6. (a) Potential energy of an electron in the Bohr orbits of an hydrogen atom. (b) Potential energy of an electron leaving a metal surface.

the other hand, fig. 2.6(b) shows the potential energy of a free electron attempting to leave the positively-charged conductor. This is compounded from an applied field of 2 V/Å and the 'image potential' which is taken to be

$$V_i = -\frac{e^2}{4x + e^2/\phi}$$

with $\phi = 4 \cdot 5 \text{ eV}$, for example. This image potential is a little arbitrary, but allows $V_i = -\phi$ for $x = 0$ (ϕ being the work function of the metal) and $V_i \rightarrow -e^2/4x$ at large x, which represents the electrostatic attraction of a charge to a plane conducting surface.

As an approximation, it may be supposed that the potential

energy diagram of the hydrogen atom approaching close to the surface is a superposition of the two diagrams, as in fig. 2.7. The extreme distortion of the potential energy surface of the atom,

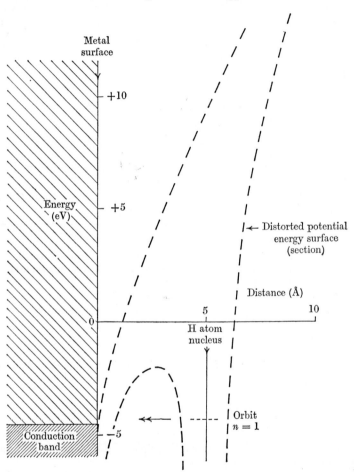

Fig. 2.7. Potential energy of an electron in field ionisation of the hydrogen atom. (The hydrogen atom is 5 Å distant from a surface of work function 4·5 eV.)

which is now apparent, means that the orbits if thought of as circular in the free atom would now be perturbed to an elliptical shape. At this point it is difficult to sustain the Bohr orbit approach to the problem, and the final understanding must come through quantum theory. When the electron in the normal ground state ($n = 1$)

2.2. Ion sources used in mass spectrometric analysis

is of the same total energy as vacant levels in the conduction band of the metal, the 'tunnel effect' allows a finite possibility of transfer of an electron (as indicated by an arrow in fig. 2.7) from the atom to the conductor through the energy barrier between the two. The fields used to produce field ionisation can obviously result in very 'thin' energy barriers, giving a correspondingly high probability of tunnelling when the atom approaches to the right distance from the surface. The positive ion so formed is immediately repelled from the surface.

It will be appreciated that this is in some respects a better-controlled mode of ionisation of molecules than impact methods, as the electron leaves the molecule with the minimum of disturbance. Certainly it is true that much less fragmentation of large molecules occurs, and that parent ions normally form the 'base peak' of the spectrum. The fragmentation that does occur can often be explained in terms of splitting of the molecular ion in the high electric field, due to polarization effects. Other minor fragmentation can arise in energetic collisions with uncharged molecules. The simplification of spectra may prove very useful in identification of substances, though isomers would be more easy to identify if some fragmentation did occur. For quantitative work, there are still troubles with the variability of the emitting element and the need for more or less frequent replacement, but improvements are continually being made. Ion currents from wire emitters are readily measurable in a mass spectrometer and are comparable with those from normal electron impact sources. Note that there is no possibility of making measurements of ionisation potential or 'appearance potential' of fragments by field emission. Ion source units have been constructed in which either method of ionisation can be switched in, to make available the best features of both types.

The spark source. This source is used for solid samples only, the sample being incorporated into one of two electrodes, between which a spark discharge is maintained under vacuum conditions. The sample material vaporises and is then ionised in the discharge. The ions are attracted out and formed into a beam, but contain a wide spread of ion energies so that a double-focusing mass spectrometer (§ 3.3) is necessary. Photographic detection is usually used to give an integrated value of the (possibly) unsteady ion current. The source has a high and approximately constant sensitivity for all elements, this being its major advantage.

The surface ionisation source. This is similar to, but not to be confused with, an arrangement described on p. 25. In this case, solid material is evaporated at a controllable rate from two relatively cool filaments (to preserve electrical symmetry) and then impinges on a much hotter filament not far away (fig. 2.8). Ionisation then occurs to an extent dependent on the difference between the ionisation potential (I) of the substance and the work function (ϕ) of the hot filament, and the temperature (T) of the latter. The proportion of particles ionised on striking hot filament is given approximately by $\exp - ([I - \phi]/kT)$. Thus the sensitivity of this method is strongly dependent on I, a fact which can be put to good use in the determination of impurities of low ionisation potential in a sample. There is also little interference from background gas, for the same reason. With the absence of magnetic field, it is favoured for accurate measurement of isotopic ratios (see p. 80).

Focused radiation or 'microprobe' sources. New techniques are presently being developed in which a polished surface of an alloy or magma sample can be minutely explored by an intensely focused beam of radiation. Different groups of workers have used variously a continuous beam of ions (e.g. Ar^+ of several keV energy and current density up to $40 \mu A/mm^2$) or light from an arc or a pulsed laser beam (0·1 joule or more per pulse). The spot may be focused to as little as 1μ diameter. The consequent evaporation of the sample produces both neutral and ionic species, and the latter may be extracted directly into the analyser and yield information on the nature of the sample at the point explored.

In one notable apparatus, an area of about $0·5 mm^2$ has been evenly 'illuminated' with a beam of ions and through correct focusing combined with magnetic dispersion of the secondary ions a magnified image of any component in the surface can be produced (Castaing & Slodzian). The resolution of detail is of the order of 1μ in these images. A finely-focused beam of ions is being used in the exploration of surfaces of metals and alloys (Liebl). The laser microprobe has been used, to quote one specific example,

2.2. Ion sources used in mass spectrometric analysis

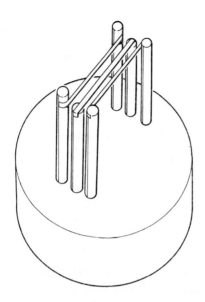

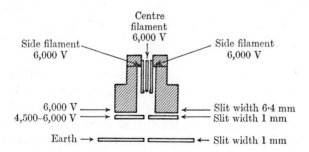

Fig. 2·8. Sketch of a triple filament assembly and section of the surface ionisation source. [From G. H. Palmer (1959). *Adv. in Mass Spectrometry*, **1**, 89. (Ed. J. D. Waldron). Pergamon.]

to study evaporation effects at different faces of a single crystal of selenium (Knox) and further developments of such probing are to be expected.

2.3. 'Natural' sources of ions. In this section a summary is given of some investigations into systems where ionised species are found more or less as a by-product of other more spectacular events. The four systems discussed are electrical discharges, flames, shock waves and the upper atmosphere (McLafferty). It may be questioned whether the adjective 'natural' may be applied to all these, but they have it in common that they were first studied for other properties than ionic content, and that studies of this last are pursued as an interesting complement to the earlier, and continuing, work on their other aspects. Mass spectrometric analysis has been applied in ways best suited to each problem in turn, to reveal something of the modes of formation and reaction of the charged species.

It is a common problem in these investigations that the sample must be taken from a region of high pressure (up to 1 atm.) and transmitted to a mass spectrometer operating at low pressure (around 10^{-5} torr) with, ideally, no distortion of the ion population.

For the sampling of ionic species the method almost invariably used is that of a 'pinhole' in a thin wall dividing the system under study from the analysing equipment. An aperture much longer than its breadth is to be avoided on account of loss of ions to the wall, or unwanted reactions on the wall. On the other hand, the more elaborate arrangements required for the production of well-collimated molecular beams (as discussed briefly on p. 91) convey little advantage over the simple pinhole for charged species.

Generally it is desirable to have as large a pinhole diameter as possible (in the thinnest possible wall), an upper limit being set for each situation by

(a) the density, and to a less extent the mean molecular weight and temperature of the gas to be sampled, these being the factors deciding the rate of ingress of gas molecules;

(b) the highest gas density which can be permitted in the region downstream of the pinhole. This is governed by the distance the ions must travel in this region, and must be such as not to allow a high probability of further reaction;

2.3. 'Natural' sources of ions

(c) the maximum 'pumping speed' which can be applied to the region discussed in (b); this is largely governed by a mixture of geometric and financial considerations.

The information relevant to the apparatus shown in fig. 2.10 is taken, to provide a numerical example. The gas density (for 1 atm. and 2000 °K) corresponds to about 120 torr at 300 °K, and the mean molecular weight will be slightly, but not seriously less than that of air. At such pressures it is known that viscous flow will occur through the pinhole since its diameter will equal many mean free paths. A useful approximate formula for the rate of intake of gas (for air at room temperature) is

$$C = 20A,$$

where A is the area of the aperture (cm²) and C the 'gas conductance' (l/s).

The distance to be travelled in the first low-pressure region is 5 cm and to ensure the fulfilment of condition (b) a pressure of 10^{-3} torr or less (when 5 cm $\equiv$ upwards of 10 mean free paths) should be maintained. The pumping speed decided by condition (c) may be taken as $S = 200$ l/s over the pressure range 10^{-5} to 10^{-3} torr. Hence for the steady state flow conditions, the maximum quantity of gas which may be allowed to flow through the system is

$$Q_{max} = 120 \times C = 10^{-3} \times S \text{ l.torr/s},$$

whence $A_{max} = 0.83 \times 10^{-4}$ cm²

and max diameter of the pinhole $= 0.1$ mm $= 100 \mu$m.

It may also be noted that such a pinhole would remove sample from the flame at a rate of approximately 1·6 ml/s. In some problems, the rate at which sample may be removed could be a limiting factor.

In cases where the system pressure is low, as in glow discharges at low pressure, the gas flow through the pinhole will be more nearly of 'molecular' type. The corresponding approximate formula for this case (for air at room temperature) is then

$$C = 12A.$$

It is appropriate to note here that 'molecular flow' sampling will usually give quite a low ion sample, and will tend to take it from the boundary layer of a system, or the 'wall sheath' of a plasma.

On the other hand it is free of a possible disadvantage of some uses of viscous flow sampling, since the ions are at once free of further collisions and under the control of the electrostatic focusing fields. In sampling by viscous flow, the adiabatic expansion of the gases, with its attendant cooling, involves many further ion-molecule collisions and may have some undesired effect on the sample before it is fully expanded into free movement (large mean free path).

It may be stated in brief conclusion that while there may be reasons to expect some distortion of results by the sampling process, indications have been obtained in most experiments of the direction and degree of the effects which do occur, and that the results presented are believed not to be vitiated by such effects.

Ions in electrical discharges. Historically, electric discharges through gases provided the earliest sources of ions for mass spectrographs, and also some of the first indications of the occurrence of ion-molecule reactions with the detection (by J. J. Thomson) of ions which could only be assigned as H_3^+, HO_2^+, H_3O^+ and N_3^+, for examples.

Since these early days, numerous investigations of ions in discharges have been carried out, using mass spectrometers of greatly varying design. A diagram of one sampling arrangement which has recently been used is shown in fig. 2.9.

In this case, the provision of a movable cathode allows the various light and dark regions of the d.c. glow discharge to be sampled. Gas pressures were in the range 0·2–1·0 torr, and the sampling hole of diameter 0·05 mm (in 0·03 mm thick glass) was thus generally smaller than the mean free path of molecules in the discharge, giving 'molecular flow' sampling. A sample of gas thus effused through the hole, into a region of much lower pressure (10^{-5} torr) and the ions were suitably accelerated and focused into the analysers. (The analysers referred to are those of a double-focusing instrument (see §3.3), which was applied to this problem because of the wide range of ion energies usually present among the ions of a discharge.)

An example of the type of result obtained, showing the variation of ion population along the discharge, is shown in fig. 2.10. The shaded regions above the curves represent the positions of the 'cathode' or 'negative' glow on the left, and two striations of the 'positive' column on the right.

2.3. 'Natural' sources of ions

It is at once seen that the negative glow contains the highest concentration of positive ions, and also that the results in the positive column correlate with the visible striated structure. More detailed study suggested that the ion ArH^+ probably arose from the reaction

$$Ar^+ + H_2 \to ArH^+ + H$$

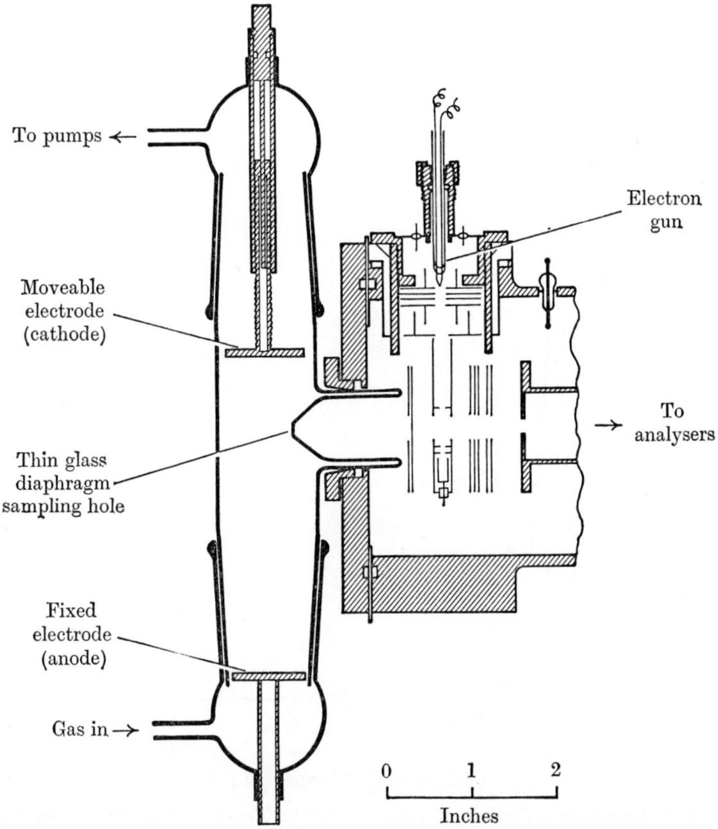

Fig. 2·9. An apparatus for sampling ions from a d.c. glow discharge. [From P. F. Knewstubb & A. W. Tickner (1962). *J. chem. Phys.* **36**, 674.]

in the negative glow, while in the positive column different conditions must result in a large contribution from some other reaction, possibly one involving argon atoms in a metastable state.

$$Ar^*_{(m)} + H \to ArH^+ + e^-$$

The molecule ion of argon almost certainly forms chiefly by a reaction involving another excited state of argon

$$\mathrm{Ar}^* + \mathrm{Ar} \to \mathrm{Ar}_2^+ + e^-$$

with a small contribution (at this pressure) from

$$\mathrm{Ar}^+ + \mathrm{Ar} + \mathrm{Ar} \to \mathrm{Ar}_2^+ + \mathrm{Ar}$$

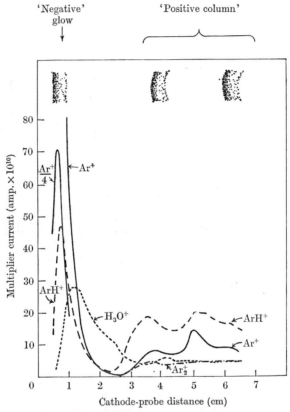

Fig. 2·10. Ions from a discharge in gas mixture 98·75 per cent argon, 1·25 per cent hydrogen. Current 0·1 mA, pressure 0·4 mm. [From P. F. Knewstubb & A. W. Tickner (1962). *J. chem. Phys.* **36**, 688.]

The H_3O^+ ion arises from tenacious traces of water impurity in the system.

As a medium for the study of ion-molecule reactions, all electric discharges have the disadvantage of the inevitable disturbing presence of electric fields, which are moreover not under complete

2.3. 'Natural' sources of ions

control, being intimately involved in the maintenance of the discharge. One may note here that interesting studies have been made of the 'afterglows' which often exist briefly after the abrupt cessation of a discharge; these afterglows can be observed under field-free conditions. Reference may also be made to the studies of the 'afterglow' when gas has passed rapidly through a discharge (see p. 50 and A.M.P.I.C. Review).

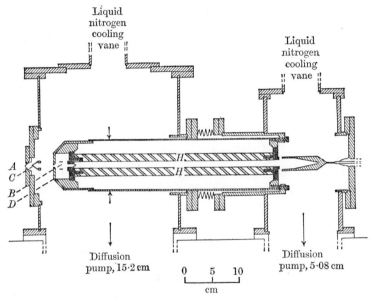

Fig. 2·11. An apparatus for sampling ions from flames at 1 atm. pressure. [From A. N. Hayhurst & T. M. Sugden (1966). *Proc. Roy. Soc.* A **293**, 36.]

Ions in flames. Various groups of workers have in the last few years investigated the ionic content of flames (usually of hydrocarbon fuels premixed with oxygen + nitrogen) burning at pressures ranging from 1 torr to 1 atm. For the study of ion-molecule reactions in the complicated régime of a flame reaction zone, the lowest pressures have the advantage of greater spatial separation of the combustion zones. The method of sampling from low pressure flames is little different from that shown in fig. 2.9. An apparatus which has been used with flames at 1 atm. pressure is shown in fig. 2.11 and has already been discussed in some detail. The pinhole sizes actually used (at A) were in the range 0·05–0·08 mm diameter. The electrodes C and B, and a negative potential on the housing D serve to accelerate and guide the ions preferentially through the various apertures. The analyser H operates at a pressure of 10^{-5} torr or less, achieved by further pumping of the second chamber as indicated. The mass analyser is of the 'quadrupole' type—see p. 69.

As a result of such experiments, the study of ionic processes in the intensely reactive primary combustion zone of such flames is seen to be, not too surprisingly, a very complicated topic. There is, however, general agreement on the main course of the reactions by which ionisation is produced and its subsequent decay. It seems that the reaction
$$CH + O \rightarrow CHO^+ + e^-$$
is the only bimolecular process involving known flame species in their ground states which could lead to ionisation; the formyl ion has been observed, though usually only in small proportion at 1 atm. pressure. It is presumed to be converted very rapidly by the proton transfer reaction.
$$CHO^+ + H_2O \rightarrow H_3O^+ + CO$$
The H_3O^+ ion is almost the only ion persisting into the burnt gases of the flame, where it decays relatively slowly (yet still at a high rate) by the recombination process.
$$H_3O^+ + e^- \rightarrow H_2O + H$$
Some measurements supporting this conclusion are shown in fig. 2.12. Many side reactions accompany this main sequence, leading to ions whose precise origins are as yet untraced.

Further examples of ion–molecule reactions can be demonstrated if metallic salts are added to the flames. In the case of lead, a rather favourable case of charge transfer seems to be observed,
$$H_3O^+ + Pb \rightarrow Pb^+ + H_2O + H$$
while with the addition of alkaline earth metals the main reaction is
$$H_3O^+ + SrO \rightarrow SrOH^+ + H_2O$$
Alkali metals ionise very readily at flame temperatures, swamping the 'natural' ionisation and showing an ionic equilibrium.
$$Na^+ + H_2O \rightleftharpoons (Na.H_2O)^+$$
The variation of this equilibrium with flame temperature has led to values of heats of hydration of such ions (e.g. 96 kJ/mole in the above case) (Hayhurst & Sugden).

2.3. 'Natural' sources of ions

While flames provide a means of studying ion-molecule reactions at high temperatures and in the presence of large free atom and radical concentrations, it is true, as for discharges, that the parameters of the system are sometimes not as freely under control as one would like. The temperature and ion generation are necessarily the joint outcome of a very vigorous chemical reaction. The most fruitful experiments have been based on hydrogen–oxygen–

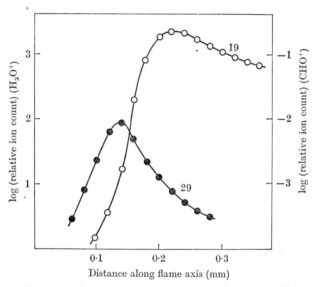

Fig. 2.12. Profiles of ion counts of mass 19(H_3O^+) and mass 29(CHO^+) through the flame front of a hydrogen flame with 1 per cent acetylene. Approximate time scale 0·1 mm = 10 μs. [From J. A. Green & T. M. Sugden, (1963) *9th International Symposium on Combustion*, Academic Press, p. 612.]

nitrogen flames (which give a high temperature, but no detectable ionisation when pure) to which small measured additions of substances are made, and the resultant ionisation studied. One may note, in passing, that the ionisation occurring in such a flame generally bears a linear relation to the amount of trace additive, and thus forms the basis of a very useful detector in gas chromatography.

Ions in shock waves. The conditions of high pressure and temperature which can suddenly be imposed on a sample of gas in a shock tube bear some resemblance to those in flames. However, it is readily appreciated that the composition of the gas mixture is much more under the control of the experimenter, and there is no necessity for the highly energetic chemical reactions which are inseparable from flame studies. Furthermore, the sudden onset of the shock heating of the sample provides an excellent 'time zero' for the study of the kinetics of subsequent processes.

It is only recently in the short history of shock wave studies that direct-sampling mass spectrometers have been applied to such studies, and after good success and great promise in the monitoring of neutral and radical species, some studies of so-called 'chemi-ions' have been made. On this topic, which is the relevant one for this section, there is perhaps still more of promise than of success, and only the oxidation of acetylene and of formaldehyde have been studied.

In the former case, an ion spectrum similar to that in low pressure flames was found, in which a number of ion types with masses from 15 to 100 a.m.u. were evident. The earliest ion in the time sequence seemed to be $C_3H_3^+$, which then rapidly disappeared if excess oxygen was present in the gas mixture, with the ion H_3O^+ rising to greatest abundance, all others remaining as minor components. The mode of formation of $C_3H_3^+$, including especially the question of its possible role as a primary ion for the system (i.e. the source of all the others), is not yet resolved. In the oxidation of formaldehyde the ion H_3CO^+ was detected early in the reaction, giving place again to the ubiquitous H_3O^+ ion (Gay *et al.*).

Ions in the upper atmosphere. The probing of conditions of the upper atmosphere by rockets and satellites is again largely a recent field and some simple and light designs of mass spectrometers have been included in the instrumentation of these studies. This has led to much interesting and useful information on the neutral composition of the atmosphere as a function of height and other parameters. Likewise the ionic content has been explored, and a fairly definite qualitative picture seems to have been produced, though much detail and explanation has yet to be worked out. The profiles of electron density against height shown in fig. 2.13 are rough averages over 'normal' conditions, and were obtained from radio observations from the ground and from measurements with electro-

2.3. 'Natural' sources of ions

static probes carried on rockets. They are included to show the scale of the effects and the nomenclature which has grown up with the subject.

Due to the incidence of a wide spectrum of far ultraviolet and X-radiation from the sun, the upper atmosphere (particularly

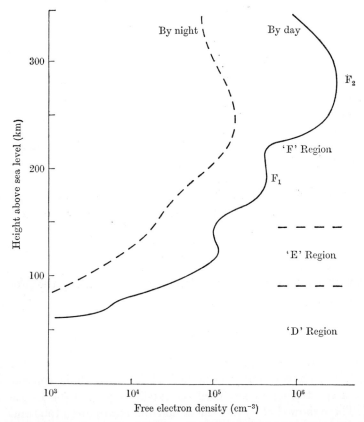

Fig. 2.13. Average profile of electron density versus height in the earth's atmosphere. [After C. O. Hines et al. (1965) Adapted from I. Paghis, T. R. Hartz & J. A. Fejer in *Physics of the Earth's Upper Atmosphere*, ed. C. O. Harris. Prentice-Hall, p. 6.]

the 'E' and 'F' layers) is the site of much photoionisation; the radiation of appropriate wavelengths is not able to penetrate to the lower regions of the atmosphere. A singular exception to this last seems to be the 'Lyman-α' line of hydrogen, which, penetrating deep into the atmosphere, may finally be absorbed by traces of NO and give rise to the NO^+ ions of the relatively low 'D' layer. In or about this region there has also been at least one report from a rocket-borne experiment of traces of other ions which appear to be H_3O^+, $H_5O_2^+$, Na^+, Ca^+, Mg^+ (Narcisi & Bailey). These could readily arise from charge transfer reactions of trace elements with NO^+, and very careful tests were made to establish that these traces did not come from the rocket exhaust gases!

In the 'E' layer (around 100 km height), the O_2^+ ion is dominant, with NO^+ also present, while in higher regions (150 km) the NO^+ ion becomes the most abundant. Above 200 km both NO^+ and O_2^+ yield first place to the ion O^+, and traces of N^+ are also found. All these observations show considerable diurnal variation, and it appears to be possible to give a satisfactory explanation of the general features in terms of the rates of photoionisation at various heights, of the recombination rates of various ions with electrons, and of one or two ion-molecule reactions of known high cross-section. Note that although N_2 is still a major component of the atmosphere at these heights, the ion N_2^+ is not found; it seems that the reaction

$$N_2^+ + O \to NO^+ + N$$

is adequate to account for this observation.

Other reactions which may be important in adjusting the proportions of ions are

$$O^+ + O_2 \to O_2^+ + O$$

and

$$O^+ + N_2 \to NO^+ + N$$

The very highest regions of the ionosphere, verging on the 'exosphere', have recently been probed for their ionic content by mass spectrometers mounted on satellites (e.g. Ariel 1—see §3.6). These show that by day the O^+ ion dominates up to 1500 km, while above this He^+ is the most abundant. At night the transition heights decrease, and above 1200 km the H^+ ion is seen to dominate over He^+. This region may be one of diffusive equilibrium modified daily by the heating effect of the solar radiation, but the control by magnetic effects becomes increasingly important at greater heights. This trend would seem to culminate in the high energy protons of the Van Allen belts, which are however somewhat beyond the bounds of mass spectrometry.

2.4. Detailed studies of ion–molecule reactions

2.4. Detailed studies of ion–molecule reactions. After the previous sections indicating that ion-molecule reactions can be from one angle an avoidable nuisance, then from another an interesting but somewhat inaccessible feature, this section attempts to show the main types of experiments which have been designed specifically for the study of these reactions. After a division of the methods into three main types, a review which attempts to be concise can hardly describe more than one or two experiments of each class in any detail, though often many variations have been used (see especially A.M.P.I.C. Review). The divisions chosen are bombardment source methods, drift methods and beam methods, with the immediate comment that these are not unique, nor even mutually exclusive classes.

Bombardment source methods. The principal method described here was one of the first used in the study of ion-molecule reactions, and has been applied by many groups of research workers. Almost all the methods use an electron bombardment source (see fig. 2.1) with small modifications to allow the deliberate use of pressures much higher than in analysis (often up to 10^{-2} torr). Ions are formed by the electron beam as before, but as these are accelerated towards the exit slit of the source, there is now a considerable chance of further collision, and of some reaction giving a different ('secondary') ion. The measured spectrum now contains some ion currents due to secondary ions (i_s) as well as the relevant primary ion (i_p) and measurement of these, with the effective length of the ion path in the source (x) and the gas pressure (n cm^{-3}) can lead to a value of cross-section for the reaction. Notice that since the reaction may occur at any time while the primary ion is accelerating from rest to an energy of several electron volts (2–10 eV), it is the experimental cross-section q_e for this range of energy which is found (see §1.2) from the expression

$$\frac{i_s}{i_p + i_s} = q_e n x. \qquad (2.1)$$

The value of q_e may be expressed as a rate constant k to a reasonable approximation by multiplying by the mean velocity of the primary ions. The acceleration of the primary ions may be varied, and the values of k extrapolated to zero (or thermal) energy; this is the value usually reported.

Several procedures are available to demonstrate that a peak in an ion spectrum arises from an ion-molecule reaction. There is first the possibility that a mass/charge ratio is observed which does not fall into the usual pattern of parent and fragment ion peaks, such as $m/e = 19$ (probably H_3O^+, if F^+ is known to be absent). This test may logically be followed by observations of the variation of the spectrum with gas pressure in the ion source. While all primary ion peaks should increase in proportion to pressure, 'secondary ions' formed by reactions in the source should show a current proportional to the square of the pressure. This amounts to a test of the above equation. Sometimes dependences of higher order are found, indicating ions formed by two or more reactive collisions in the source. A further diagnostic test may be made by altering the electric field in the ion source by varying the repeller potential. This will vary the collection efficiency of all ions, but as the field increases, the proportion of secondary ions will decrease due to the shorter residence time in the source; two distinct types of curve are therefore obtained.

Having proved the formation of a secondary ion, one must determine from which primary ion it has arisen, since in most cases several of the latter will be present. Three methods seem to be in general use:

(a) The electron energy is progressively reduced while the ratio of secondary to supposed primary ion current is noted. For a correct assignment the ratio should remain approximately constant. For example, in deciding between the reactions

$$Ar^+ + H_2 \rightarrow ArH^+ + H$$

or $$H_2^+ + Ar \rightarrow ArH^+ + H$$

the predominant occurrence of the first reaction is confirmed by constancy of i_{ArH^+}/i_{Ar^+} over a range of electron energies above the ionisation potential of Ar, while $i_{ArH^+}/i_{H_2^+}$ varies widely. Obviously this procedure will not discriminate well among primary ions of very similar appearance potential.

(b) In some cases it is possible to add a small amount of a second component to the source gas which enhances the suspected primary ion, while leaving the number density of the suspected neutral reactant unchanged. For example, in a test of whether the process

$$C_2H_4^+ + C_2H_4 - C_3H_5^+ + CH_3$$

2.4. Detailed studies of ion–molecule reactions

occurs in ethylene, ethane (which gives a large contribution of fragment ion $C_2H_4^+$) was added and was indeed found to enhance the $C_3H_5^+$ ion current (see McLafferty, p. 83).

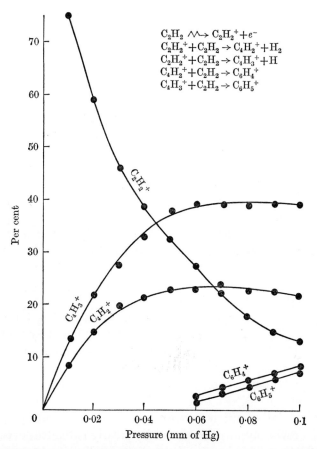

Fig. 2.14. The per cent abundance of various ionic species from acetylene as a function of pressure in the ionisation chamber. [From C. E. Melton (1963) in *Mass Spectrometry of Organic Ions* (Ed. F. W. McLafferty) Academic Press, p. 84. Originally from P. S. Rudolph & C. E. Melton (1959) *J. phys. Chem.* **63**, 916, at the Oak Ridge National Laboratory.]

(c) The various ion currents are expressed as a percentage of the total, as a function of source gas pressure. Any variation of secondary ion current should be accompanied by an opposite variation of abundance of its primary ion (see fig. 2.14).

Elaborations of the bombardment source method. Brief mention must be made of three directions in which the method described above has been extended or adapted. Some investigations have embraced more than one of these in various combinations.

(a) The bombarding particle need not be an electron, some ion sources having incorporated a radioactive element, for example coated onto the repeller electrode, which provides effective α-particle bombardment of the sample up to source pressures of several torr. Similarly, photon bombardment can be used, and in some very interesting experiments positive ions have been directed through a slightly modified bombardment source. In such experiments (almost to be placed amongst beam methods) the energy of the bombarding ions is usually known, and the ion species may also be selected by a 'source mass spectrometer'.

(b) The disadvantage, or at least difficulty, of the basic method in producing a cross-section somehow averaged over a range of primary energies has been cleverly alleviated by 'pulsed' operation of the source. In one variation, a very short burst of electrons is admitted to the source, which then remains field-free for a measured time (several micro-seconds before a very strong repeller field is applied to remove all ions from the source. The major part of any reaction occurs during the field-free period, with positive ions of thermal energy. In another recent investigation, the very short (2×10^{-8} s) electron pulse is immediately followed by another very short (5×10^{-8} s) pulse on the repeller. The primary ions are thus given a measured impulse, and travel across the source with constant and known energy. Such experiments are capable of yielding values of 'q' as a function of ion energy.

(c) By special attention to source design, the pressure limit of 10^{-2} torr previously mentioned can be greatly extended, even as high as 2·0 torr. This has the immediate implication that many collisions can occur before the ion leaves the source (making the situation akin to drift methods). The spread of primary ion energies is less, though less certainly defined, and reactions of order higher than second are likely to occur, leading to the observation of ions of high mass/charge ratio from ionic polymerisation or 'clustering' reactions.

An interesting variation (Munson and Field) uses methane at a pressure of about 2·0 torr, to which traces of other compounds

2.4. Detailed studies of ion–molecule reactions

are added. The dominant ion under these conditions is CH_5^+, which is then able to donate a proton to the trace component. In most cases this results in some degree of fragmentation of that substance, yielding an ion spectrum very similar in character to an electron bombardment 'cracking pattern', but quite different in detail. The process has been named 'chemical ionisation' by its originator, F. H. Field (though surely it would have been Field ionisation but for the sad coincidence of name!) and could prove to be an application of ion-molecule reactions which is useful to the analyst.

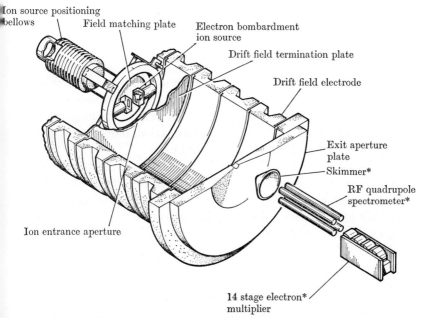

Fig. 2.15. Drift tube with mass spectrometer.

* These components are discussed in other sections of the book. [From D. L. Albritton et al. 1968. Phys. Rev. **171**, 95.]

Drift methods. The essential feature leading to the term 'drift' is that a cloud of ions moves slowly through a gas, with its thermal velocity distribution barely disturbed. This régime has been studied for many years in apparatus where the 'drift' velocity is determined by a weak electric field.

This method has been very highly developed and a recent apparatus is illustrated in fig. 2.15. In this apparatus, a type of electron-bombardment source generates ions which pass through an 'entrance aperture' and immediately settle to a steady drift motion down the tube, at the end of which a sample enters the mass spectrometer for analysis. The electron source can be pulsed, and the time for ions to travel down the tube measured; this time can be varied widely by altering the gas pressure, the electric field and the distance of the source from the exit aperture. The extraction of a rate constant from the data is not simple, and requires knowledge of diffusion coefficients and mobilities of the ions concerned, as well as an analysis of their interplay with the ion-molecule reactions under study. The reaction so far given most attention is of the clustering type in nitrogen.

$$N_2^+ + 2N_2 \rightleftharpoons N_4^+ + N_2$$

In another recent variation of the method, the electron bombardment source is replaced by a complete mass spectrometer which can then inject a selected ion species into the drift chamber. This now gives certainty of the primary ion, and an independent choice of the reacting molecule (Kaneko et al.).

Close in concept to the electrical drift apparatus are the more recent types of experiment in which the whole sample of gas has a certain drift velocity, superimposed on the purely thermal motions of both ions and neutral species. In one such experiment a rapidly flowing gas stream is crossed by a beam of photons of high energy (fig. 2.16). The distance between the point of formation of the ions and the sampling point is not variable in this specific case, but flow velocity and pressure are useful variables. Reaction times of 25–100 μs at pressures in the range 0·2–2·0 torr are typical, and again clustering reactions such as the one just mentioned are the subject of study.

A very adaptable arrangement is provided by the recent 'flowing afterglow' technique, illustrated in fig. 2.17. In this case, the gas flow, usually of helium, passes through an electrical discharge and is partially ionised. The time of travel to the mass spectrometer inlet is about 6 ms, and the pressure around 0·3 torr. The reaction
$$He^+ + 2He \rightarrow He_2^+ + He$$
is observed and measured.

The extreme flexibility of this apparatus becomes evident

2.4. Detailed studies of ion–molecule reactions

when 'chemical manipulation' is used. The addition of a slow flow of (e.g.) nitrogen after the discharge results in formation of atomic and molecular ions of nitrogen by reactions with the helium ions

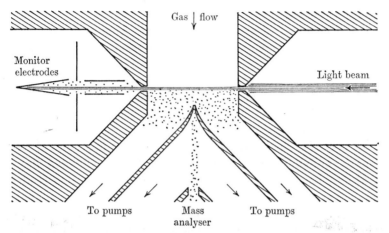

Fig. 2.16. Reactions in a rapid gas flow.

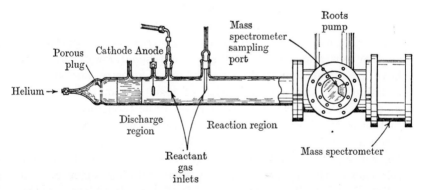

Fig. 2.17. The flowing afterglow reaction system of Fehsenfeld, Schmeltekopf and Ferguson. [From F. C. Fehsenfeld et al. (1966). J. chem. Phys. **44**, 3022.]

and with the metastable excited helium atoms, which are produced in the discharge and survive into the afterglow.

$$He^+ + N_2 \rightarrow N^+ + N + He,$$
$$He(2^3S) + N_2 \rightarrow N_2^+ + He + e^-$$

It is possible with suitable control to treat these nitrogen ions

as the primary ions for a reaction with a second gas, added at the second inlet, e.g.
$$N_2^+ + NO \to NO^+ + N_2$$
Similarly the reaction
$$O^+ + CO_2 \to O_2^+ + CO$$
has been studied.

The calculation of exact rate constants from such an experiment is again far from easy, involving diffusion and recombination effects which may vary between the different ions. However, within an accuracy of 20–30 per cent a great deal of useful information can be obtained in a field where so much is still completely unmeasured (see A.M.P.I.C. Review).

Beam methods. The highest level of elaboration in studies of reactions is probably reached in methods using collimated beams of atoms, ions or molecules. For ion–molecule reaction studies, the system most often found will produce a beam of ions of known energy (or speed) and direction. Often, though not always, the primary ion species is selected by a source mass spectrometer. The beam is passed through a collision region where the reactant molecule is present at an elevated pressure, and some reaction will occur (see, e.g. fig. 1.2). Finally, the fullest analysis of the products would include both velocity and mass analysis carried out as functions of angular separation from the primary beam. So far, few experiments have investigated more than two of mass, velocity, and angular distributions, and a combination of results from two or more moderately complicated experiments may be thought easier than further elaboration of apparatus. One very complete example of a beam apparatus is shown in fig. 2.18, where one may trace the beam-forming sections followed by mass analysis and velocity selection of the primary ions. The small collision region is then followed by velocity and mass analysis of all resultant ions, and the apparatus doing this can be pivoted on an axis coincident with the collision region, to explore angular distributions.

Two further variations may be noted among beam methods. The double or crossed-beam experiment is a final refinement which can define most closely the encounter of the reacting particles. The neutral reactant does not simply fill the collision regions as a randomly moving gas, but is itself formed into a molecular beam, of which the direction and possibly the velocity distribution are

2.4. Detailed studies of ion–molecule reactions

known. The effective collision region is where these beams intersect. A diagram of such an apparatus is shown in fig. 2.19. This method is necessarily used when the neutral reactant is an unstable atom or radical (e.g. H, OH).

One type of reaction which has received particular study is the charge-transfer (see p. 55). Since this occurs by the transfer of

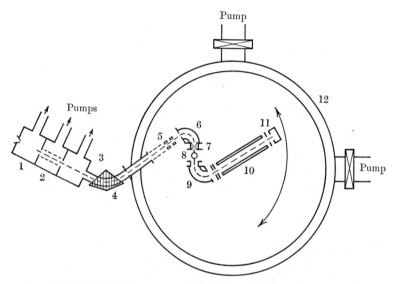

Fig. 2.18. A single-beam instrument with measurement of energy and angular distribution. 1, Ion source; 2, focusing region; 3, mass selector; 4, magnet; 5, primary ion retardation and focusing; 6, 7, velocity selector*; 8, collision region; 9, velocity analyser*; 10, quadrupole mass analyser*; 11, ion multiplier; 12, main vacuum chamber.

* These components are discussed in other sections of the book.
[From R. L. Champion et al. (1966) J. chem. Phys. **45**, 4377.]

an electron from one particle to the other, very little momentum is transferred. In this case, velocity analysis of the secondaries is meaningless, and they will have to be given some drift velocity or impulse to make them leave the collision region.

Provided that the effective dimensions of the collision region are known, and also the particle density within it, the derivation of microscopic cross-section q as a function of ion energy follows fairly simply, and has in some cases been plotted to ion energies as low as 1 eV. However, the full detail of information from a beam experiment can give further information about the reaction. It is possible for a reaction to proceed through formation of a complex, in which all the atoms present play some part, or through a so-called 'stripping' mechanism in which only one atom of either the target or the projectile molecule is involved; the others have been described as 'spectators' of the event. A comprehensive example of the possibilities is provided by a reaction such as

$$CH_4^+ + CH_4 \rightarrow CH_5^+ + CH_3$$

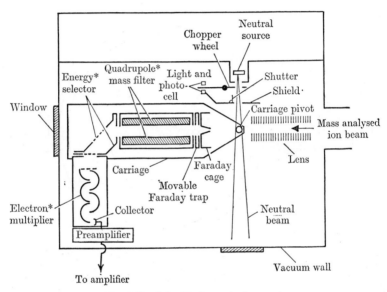

Fig. 2.19. A double-beam instrument.

* These components are discussed in other sections of the book.
[From B. R. Turner *et al.* (1965). *J. chem. Phys.* **42**, 4089.]

The mechanism of this might be characterised by comparing the velocity of the CH_5^+ ion with that of the projectile CH_4^+ ion. If a 'complex' mechanism is involved, in which the ion $C_2H_8^+$ has a transitory existence, the momentum would be fully shared and the velocity ratio would be half. If the mechanism were a 'stripping' of a hydrogen atom by the projectile, the velocity ratio would be 16/17, while a proton transfer reaction would yield a ratio of 1/17. These exact ratios would possibly be modified by thermal motions of the neutral reactant, and also by any inelastic nature

2.4. *Detailed studies of ion–molecule reactions* 55

of the collisions. The formation of a complex has been sub-divided according to whether the initial relative kinetic energy is more or less completely absorbed into internal energy (an inelastic collision complex) or re-appears completely as kinetic energy of products (an elastic collision complex). Theory may be applied to calculate angular distributions of products for each model, and these may be appreciably different. However, it will be realized that each represents an extreme case, and in comparing with experiment it is likely that one can only say to which model an actual reaction is nearest. Generally, complex formation (akin to the strong coupling of §1.7) is likely at low relative energies, and stripping reactions are more likely at high energies.

2.5. Classification of ion–molecule reactions. The somewhat bewildering array of reactions which may occur in encounters of an ion and a molecule may be subjected to some degree of classification on the following lines. The semi-pictorial symbols are the author's attempt at condensation of the essential features of each class, and show the entity which is apparently involved, and its direction of transfer when the reactant ion is written first.

Transfer of an electron.

(a) The positive ion may abstract an electron to give simple charge transfer $\overset{\frown}{e^-}$ $AB^+ + CD \rightarrow AB + CD^+$

With simple substances, these reactions have received considerable study using high energy ion beams (50–5000 eV). More recently, the energy range has been extended downwards to about 5 eV when it is often found that other types of ion molecule reaction begin to compete. Cross-sections for charge exchange usually decrease towards low energies, except for the case represented as

$$AB^+ + AB \rightarrow AB + AB^+$$

which is termed 'resonant charge transfer', and usually shows the contrary behaviour, with quite large cross sections (approximately $10-100 \times 10^{-16}$ cm^2) at low ion energies. Interesting examples are

$$Ar^+ + N_2 \rightarrow Ar + N_2^+$$
$$H_3O^+ + Pb \rightarrow Pb^+ + H_2O + H$$
$$Ar^+ + CH_4 \rightarrow Ar + CH_4^+$$

Charge exchange reactions can be responsible for considerable attenuation of ion beams in a mass-spectrometer, if the degree of vacuum is not as good as it should be.

(b) Charge transfer may be followed by dissociation due to the excess energy of the newly-formed ion. (Any dissociation of the fast neutral molecule is generally not observable with a mass-spectrometer.) $\overset{\frown}{e^-}*$ $AB^+ + CD \rightarrow C^+ + D + AB$

Such effects as this can be observed in a mass-spectrometer when mixtures of gases are admitted. For example, in a 10:1 mixture of argon and methane at a source pressure of up to 10^{-6} torr the ions which are seen arise from the processes

$$Ar + e^- \rightarrow Ar^+ + 2e^-$$
$$CH_4 + e^- \rightarrow CH_4^+ + 2e^-$$
$$\rightarrow CH_3^+ + H + 2e^-$$
$$\rightarrow CH_2^+ + H_2 + 2e^-$$
$$\rightarrow CH^+ + H + H_2 + 2e^-$$

The array of peaks resulting from the methane is simply its 'cracking pattern' for the conditions adopted, and all the peak height ratios are predictable from standard calibration experiments. If the source gas pressure is now raised to about 10^{-4} torr, it is found that the anticipated peak height ratios are not obtained. This is evidently due to the enhanced effects of the charge transfer reactions

$$Ar^+ + CH_4 \rightarrow CH_3^+ + H + Ar$$
$$\rightarrow CH_2^+ + H_2 + Ar$$

which must therefore have quite large reaction cross-sections. Although such effects can be avoided by the use of lower source pressures, their study is important for an understanding of radiolysis of gases at high pressures.

Dissociative charge transfer has also been employed as a means of adding a well-defined quantity of energy to an ion (most frequently a hydrocarbon species) and studying its decomposition. The results are of interest in the detailed understanding of mass spectra (see chapter 5).

2.5. Classification of ion–molecule reactions

Transfer of heavy ions.

(a) A positive ion may donate a proton (or in some cases acquire a hydride ion H⁻)

$$\overset{\frown}{H^+} \quad ABH^+ + CD \rightarrow CDH^+ + AB$$
$$\overset{\frown}{H^-} \quad AB^+ + CDH \rightarrow CD^+ + ABH$$

Such proton transfer processes are quite common and usually of high cross section (10^{-14} cm²) while the hydride ion transfer shows somewhat lower values. Examples of such reactions are

$$\overset{\frown}{H^+} \quad ArH^+ + H_2O \rightarrow H_3O^+ + Ar$$
$$CHO^+ + H_2O \rightarrow H_3O^+ + CO$$
$$H_3O^+ + SrO \rightarrow SrHO^+ + H_2O$$
$$\overset{\frown}{H^-} \quad CH_3^+ + C_2H_6 \rightarrow C_2H_5^+ + CH_4$$

Series of such reactions have been studied with a view to comparing proton affinities of different species, and to finding approximate values (see p. 88).

(b) The proton or hydride ion transfer may be followed by dissociation of the new ion

$$\overset{\frown}{H^+} \quad {}^*ABH^+ + CD \rightarrow CH^+ + D + AB$$
$$\overset{\frown}{H^-}{}^* \quad AB^+ + CDH \rightarrow C^+ + D + ABH$$

This variation of proton transfer has recently received close attention in the experiments called 'chemical ionisation' mass spectrometry (see p. 49). Other experiments (as mentioned under dissociative charge transfer) use these reactions too, and will be discussed in chapter 5.

(c) In rarer cases, other charged groups (e.g. I⁺) are thought to be transferred in a reaction. For example,

$$HI^+ + CH_3I \rightarrow CH_3I_2^+ + H$$

Transfer of neutral groupings. The transfer of a chemical group or atom X can in principle be either from or to the ion, though cases of the former seem rare

$$\overset{\frown}{X} \quad \text{(rarely)} \quad ABX^+ + CD \rightarrow AB^+ + CDX$$
$$\overset{\frown}{X} \quad \quad AB^+ + CDX \rightarrow ABX^+ + CD$$

This is the widest class of reaction, and while X is most frequently the hydrogen atom, several other atoms or simple chemical groups have been observed to be transferred, for example, N, O, I, Cl, Br, HI, CH_3, CH_2, C_2H. Sometimes it seems that an exchange of groups or atoms must occur (a replacement reaction). Cross-sections for this class cover the whole range of values. Examples which have already been cited in this book are

$\overset{\frown}{O}$ $N^+ + O_2 \to NO^+ + O$

$\overset{\frown}{N}$ $O^+ + N_2 \to NO^+ + N$

$\overset{\frown}{O}$ $O^+ + CO_2 \to O_2^+ + CO$

$\overset{\frown}{H}$ $Ar^+ + H_2 \to ArH^+ + H$

The above classes do not form an unambiguous system, even if both reactants and products of a reaction are known. For examples one may mention

$\overset{\frown}{H^+}$ or $\overset{\frown}{H}$ $CH_4^+ + CH_4 \to CH_5^+ + CH_3$

$\overset{\frown}{H^+}$ or $\overset{\frown}{H}$ $H_2^+ + H_2 \to H_3^+ + H$

$\overset{\frown}{CH^+}$ or $\overset{\frown}{CH}$ $C_2H_4^+ + C_2H_4 \to C_3H_5^+ + CH_3$

$\overset{\frown}{N^+}$ or $\overset{\frown}{O\,N}$ $N_2^+ + O \to NO^+ + N$
replacement

$\overset{\frown}{e}$ or $\overset{\frown}{O}$ $O^+ + O_2 \to O_2^+ + O$

Perhaps the fairest comment on such ambiguities is that ion–molecule reactions involving transfer of heavy groupings will, under many experimental conditions, proceed through formation of a complex. The classification is then little more than a convenience.

Attachment reactions. The attachment of a complete molecule to an ion is always a much less rapid process than any of the above, and while there is evidence in isolated cases to suggest a two-body process, in others it seems much more likely to occur in a three-body collision. As explained in §1.4, the time of an orbiting type of ion-molecule collision may be quite long. There is considerable evidence for the occurrence of these 'sticky collisions', perhaps even

2.5. Classification of ion–molecule reactions

meriting the name of unstable complexes, which if not stabilised by further collision may decompose in one or more ways. For example,

$$C_2H_4^+ + C_2H_4 \to [C_4H_8^+]$$

$$[C_4H_8^+] \begin{cases} \to C_4H_7^+ + H \\ \to C_3H_5^+ + CH_3 \end{cases}$$

In this specific case, the postulated intermediate ion has in fact been detected. Other systems in which attachment reactions have been observed to give stable products are (omitting mention of third bodies)

$$N_2^+ + N_2 \to N_4^+$$
$$Ar^+ + Ar \to Ar_2^+$$
$$C_2H_2^+ + C_2H_2 \to C_4H_4^+$$
$$Xe^+ + (CN)_2 \to Xe(CN)_2^+$$
$$C_2H_5I^+ + C_2H_5I \to C_4H_{10}I_2^+$$
$$O_2^+ + O_2 \to O_4^+$$
$$NH_4^+ + NH_3 \to N_2H_7^+$$
$$H_3O^+ + H_2O \to H_5O_2^+$$
$$Na^+ + H_2O \to NaH_2O^+$$

It is a point of interest to note that with such weakly bound product ions, as some of these are, the reverse reactions may be observable, and conditions even of equilibrium may be attained in some situations leading to observation of such ions as

$$O_6^+, \quad O_8^+, \quad N_3H_{10}^+, \quad H_7O_3^+, \quad H_9O_4^+, \quad Na.2H_2O^+$$

Collisions of high energy ions (energy > 1 keV). These events may result in ionisation of the target molecule, sometimes with dissociation. They may occur in a mass spectrometer flight tube, but are more relevant to studies in radiation chemistry, and are somewhat on the fringe of the topic of ion–molecule reactions.

Reactions of negative ions. Relatively few reactions involving negative ions have been found to occur, and they can be fitted into the classification given. However, they are rather less well established, and the subject of negative ions is not given detailed attention here.

3 Types of mass spectrometer

3.1. Introduction. From the historical aspect, the first devices for the measurement of mass/charge ratios of atomic and molecular particles were the parabola mass spectrograph of J. J. Thomson in 1910 and the 'velocity focusing' mass spectrograph of F. W. Aston in 1919. These designs are no longer in use, and the interested reader is referred to fuller treatises.

The subsequent developments in this field have produced a wide variety of 'mass analysers', some employing magnetic fields and some not. Direct and alternating electric fields are used in various combinations. The aim of this chapter is to outline the principal forms now in use, and to suggest their capabilities and limitations, without great detail of individual design.

Amongst the many important characteristics of the instruments is the 'resolving power' or 'resolution', expressing the ability to separate ions of adjacent mass numbers. This is calculated in terms of the apparent width (ΔM) of the peak which appears for one type of ion (mass M), when plotted as ion current versus mass number.

$$\text{Resolution} = \frac{M}{\Delta M}$$

Unfortunately the exact conditions for measurement of ΔM are not yet standard, and it may refer variously to points of 50–1 per cent of maximum peak height. Alternatively it may sometimes be expressed as the minimum ion current between two peaks of equal height and adjacent mass number (the 'per cent valley' definition). As a comparison of ability here, approximate figures will be given for each type which are intended to represent measurement of ΔM at 5 per cent peak height ('10 per cent valley').

3.2. Single-focusing instruments. The earliest design of instrument which is still met with today is the '180° deflection' mass spectrometer, first built by Dempster. The salient features of this

3.2. Single-focusing instruments

are the generation of an ion beam which is as far as possible homogenerous in energy, followed by deflection of this through 180° in a uniform magnetic field, as seen in fig. 3.1. The motion of the ions is governed by the force law

$$Bev = \frac{mv^2}{r}$$

i.e.
$$\frac{m}{e} \cdot v = Br \qquad (3.1)$$

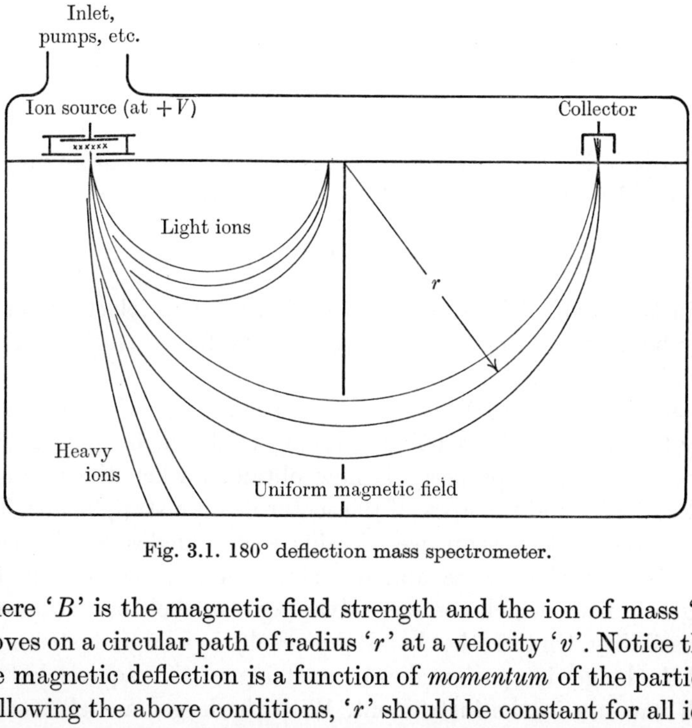

Fig. 3.1. 180° deflection mass spectrometer.

where 'B' is the magnetic field strength and the ion of mass 'm' moves on a circular path of radius 'r' at a velocity 'v'. Notice that the magnetic deflection is a function of *momentum* of the particle. Following the above conditions, 'r' should be constant for all ions of one species. It may then be seen that ions commencing from a given point, in a somewhat divergent beam, will reach a point of best (though not perfect) re-focus after deflection through an average of 180°. This 'direction focusing' is the property referred to in the heading of this section.

Instruments of this type can be made very compactly, usually with a permanent magnet, and are in use for analysis of light gases, or residual gases in a vacuum system. The resolution achieved in small designs will rarely exceed 50, but this has been increased to as much as 400 in larger versions. The limits are the accuracy of construction, and uniformity of magnetic field and of beam energy. The latter is dependent on producing ions at thermal energies (see §2.2) and accelerating these through a potential (V) of 100–1000 V. The spread of energies due to initial velocities is then relatively small, and we may write

$$eV = \tfrac{1}{2}mv^2.$$

Hence
$$v = \frac{(2eV)^{\frac{1}{2}}}{(m)} \tag{3.2}$$

and from the previous result it follows that

$$\frac{m}{e} = \frac{B^2 r^2}{2V} \tag{3.3}$$

Using practical units of gauss, cm and volts, the condition may be expressed for a singly-charged ion in the form

$$M = \frac{B^2 r^2}{V} \times 4{\cdot}82 \times 10^{-5} \text{ at. mass units} \tag{3.4}$$

Adjustment of 'V' (or 'B' for a design with an electromagnet) may be used to bring 'r' to the right value for measuring the desired ion current.

Variations on the Dempster design use a magnetic sector with only 90°, or sometimes 60° deflection, as shown in fig. 3.2. The direction focusing property is now obtained when the entrance and exit slits and the apex of the sector lie on a straight line. The most important consequence of using sector angles smaller than 180° is that the source and detector can now be outside the magnetic field, which conveys advantages in their operation and accessibility. On the other hand, this results in a larger instrument, and the longer ion path makes more stringent the requirement for high vacuum in the instrument. With the larger models usually constructed, resolution in the range 500–1000 is possible.

3.3. Double-focusing instruments. The introduction of instruments which would bring to a focus ions which initially varied both in direction and energy (hence double-focusing) was spurred by the necessity, for some purposes, of obtaining good resolution even with such ion sources as a spark, from which ions emerge with a range of energies. The earliest work of this type was directed towards the detection of isotopes in a wide range of elements, with

3.3. Double-focusing instruments

accurate measurements of their relative abundances and masses. The modern design of one such instrument, constructed as a mass-spectrograph in which a complete mass spectrum may be recorded on a photographic plate, is shown in fig. 3.3 (a). Another design, shown in fig. 3.3(b) is used for peak by peak measurement (correctly, mass spectrometry) at high resolution.

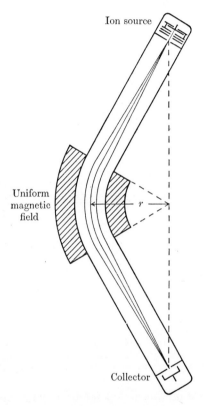

Fig. 3.2. 60° deflection mass spectrometer.

Both designs follow the general pattern of passing the ions first through the electric sector field (an energy analyser), producing dispersion according to energy, and then through a magnetic sector field, producing dispersion according to momentum. By the use of 'ion optics', it can be arranged that the 'chromatic aberrations' of the two analysers, considered as ion lenses, cancel

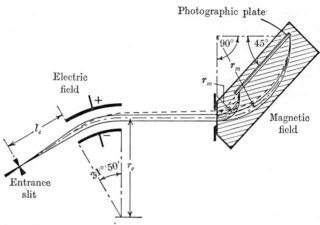

Fig. 3.3(a). The Mattauch–Herzog double-focusing mass spectrometer. ---- shows a trajectory of ions of somewhat increased energy. [After J. H. Beynon (1960). *Mass Spectrometry and its Application to Organic Chemistry*, Elsevier.]

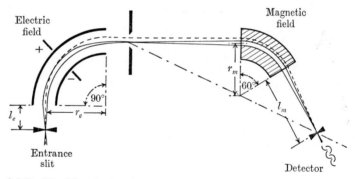

Fig. 3.3(b). The Nier–Jordan double-focusing mass spectrometer. ---- shows a trajectory of ions of somewhat increased energy. [After J. H. Beynon (1960). *Mass Spectrometry and its Applications to Organic Chemistry*, Elsevier.]

out. Thus a focus is produced dependent only on the mass (strictly, the mass/charge ratio) of the ions.

The motion of the ion in the electric sector field is of course governed by the force equation

$$eX = \frac{mv^2}{r_e} \tag{3.5}$$

Hence the energy of the ion is given by

$$eV = \tfrac{1}{2}mv^2 = \frac{r_e eX}{2} \tag{3.6}$$

3.3. Double-focusing instruments

As may be imagined, instruments of this class tend to be large and expensive, but are capable of resolution as high as 100,000. At such figures, the deviation of atomic and molecular masses from integral values are accurately measurable, and peaks which appear simple under low resolution may be revealed as multiplets. The accurate, absolute measurement of molecular weight has important applications in qualitative analysis of large molecules (see p. 77).

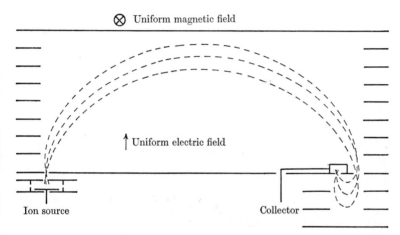

Fig. 3.4. A cycloidal mass spectrometer. The drawing illustrates the direction-focusing property for ions of the same energy at the entrance slit.

It should not be imagined that the sector field instruments are the only ones providing double focusing action. The cycloidal mass spectrometer (fig. 3.4) is a remarkable example of a compact instrument with this capability, yet it does not seem to have received as wide an application.

Uniform magnetic and electric fields are maintained, at right angles to each other, throughout the volume traversed by the ions. The ions need only enter at a slit with an approximately defined energy and direction. They then travel in cycloidal (sometimes called trochoidal) orbits, returning in theory to a perfect focus at the collector slit. A resolution of about 4,000 has been claimed for one such instrument.

3.4. Time-of-flight mass spectrometer.

The time-of-flight mass spectrometer is in a class by itself in its use of pulsed ion beams as an essential feature. The operation consists of the formation of a sample of ions in a well-defined position (usually by a brief pulse of electrons of suitable energy) followed by a rapid acceleration of the ions into the flight tube by sudden application of an axial electric field. The aim is to give all ions the same energy,

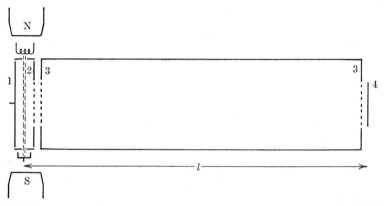

Fig. 3.5. Time-of-flight mass spectrometer. 1, Electron impact ion source (as in fig. 2.1); 2, draw-out grid (pulsed); 3, flight tube ($-3{,}000$ V); 4, cathode of multiplier detector (special design).

so that their velocities, and hence the times taken to reach the detector, are a function of the mass/charge ratio. Referring to fig. 3.5, the effective length of the flight path is l, and the ions fall through a potential V when they are launched from the initial position.

Hence

$$\text{energy of ions} = \tfrac{1}{2}mv^2 = eV$$

$$\text{time of flight} = l/v = l\left[\frac{2eV}{m}\right]^{\frac{1}{2}}.$$

The complete mass spectrum of the ions in the pulse, or parts of the spectrum, may be displayed directly on an oscilloscope for examination and a photographic record. The complete mass spectrum can thus be obtained and displayed in less than 40 μs, and the cycle is repeated at least 10,000 times/s. It may readily be appreciated that this instrument is uniquely suited to the study of rapid reactions provided that the sample can be conducted to the

3.4. Time-of-flight mass spectrometer

source in a suitably direct manner. Although it is true that there is an adverse factor concerning sensitivity in comparison with other instruments, because in the most common mode of operation the source is only active in generating ions for less than 1 per cent of the time, it is on the other hand a considerable compensation that no narrow slits or apertures are needed to define a beam. It is certainly unrivalled by magnetic spectrometers in maintaining its resolution and sensitivity even on single scans with a repetition interval of 50 μs. A very good example of this is its use in the study of shock wave phenomena (see p. 42).

The instrument is also widely used with less rapidly changing systems and even for static analysis, when the rapidity of oscilloscope presentation is less important than an electronic gating system which allows accurate spectrometry of more conventional type. This operates by collecting separately the ions (in fact, electrons derived from the ions by multiplication) which arrive during an interval of only 5×10^{-8} s on each cycle. The timing of this interval may be adjusted to coincide with that of arrival of any desired ion, and thus a current is measured representing its abundance in the spectrum. Alternatively the timing may be varied slowly and regularly, when the current measurement will yield a mass spectrum in exactly the same way as the scanning of a magnetic spectrometer. A further unique feature of the system is the possibility of using up to six channels of measurement simultaneously by multiplication of the appropriate circuitry. A resolution of up to 600 may be maintained in the best circumstances.

3.5. Mass-spectrometers employing radio-frequency fields.
The linear accelerator type. This bears a superficial resemblance to a time-of-flight spectrometer but differs in its operation, and moreover has generally been constructed as a much smaller and less ambitious instrument. It requires the entry of ions at a certain energy, so that different mass/charge ratios are represented by different velocities. The ions then pass through a system of grids grouped in threes, the central grid of each group being connected to a r.-f. supply. Ions gain or lose energy in each group according to the phase of the r.-f. field as they cross the gaps. Only for the mass/charge ratio to which the array is 'tuned' will some ions (those entering in favourable phase) receive an increase in energy at each

gap, and only these will be able to pass the final energy barrier (retarding grid) and reach the detector. The system may be pictured as analogous to a series of synchronised traffic lights. Different ions are detected by variation of the radio-frequency.

Instruments of this type commonly show resolution up to 50, but occasionally as high as 250. It has been possible to make them in very light and compact form, which has found application in space and upper atmosphere probes.

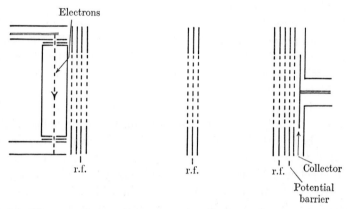

Fig. 3.6. Linear accelerator (Bennett-type) radio-frequency mass spectrometer. [After J. H. Beynon (1960). *Mass Spectrometry and its Application to Organic Chemistry*, Elsevier.]

The omegatron. This device employs a magnetic field crossed with a radio frequency which is varied to scan the mass spectrum—hence the name.

As indicated in the diagram (fig. 3.7) the ions which are 'on tune' are driven in an expanding spiral, gaining energy in each half-revolution as in the operation of a cyclotron. Other ions gain energy for only a few cycles, then by getting out of favourable phase with the r.-f. field, loose energy and spiral inwards. Thus only the desired ions are collected at the outside of the spiral. The others must drift slowly along the magnetic field axis and be lost at the end plates.

The device can be made very compact, and has high sensitivity, due to the high efficiency of collection of all the ions of correct mass/charge formed on the axis. As in other instruments, a narrow beam

3.5. Mass spectrometers employing radio-frequency fields

of electrons is the usual ionising agent. The energy of ions along the magnetic field axis does not directly affect the operation, but it must be low—hence it is difficult to introduce ions as such from outside the instrument. One definite limitation of operations is the requirement of quite high vacuum conditions, due to the long ion path; residual gas analysis in high vacuum is thus the commonest use of the omegatron.

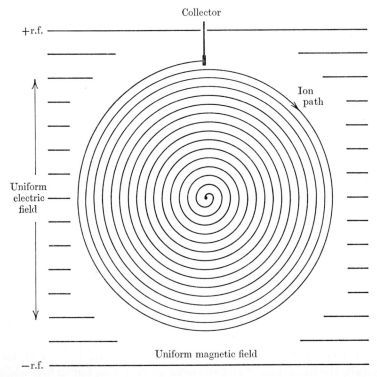

Fig. 3.7. Omegatron mass spectrometer. The uniform electric field is maintained by a series of 'picture frame' plates. The resolution is proportional to the length of the ion path; the drawing corresponds to a resolution of about 23.

The resolution varies inversely with mass number and r.-f. voltage, but can be as high as 1000 at low masses; its use is generally confined to the measurement of masses below 50 a.m.u.

The quadrupole mass-spectrometer. Often, and perhaps more correctly, referred to as a mass filter, this instrument has a simple construction but a rather complicated analysis of operation. If the x and y directions of fig. 3.8 are considered, the mean potentials form respectively a well and a hill, as shown in fig. 3.9. In the x direction, the ions have natural frequencies of oscillation in the well, which are generally considerably lower than the applied frequency, but highest for the lightest ions. The forced oscillation reaches a maximum amplitude which is higher the lighter the ion, and hence light

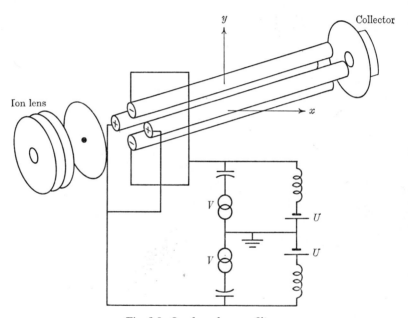

Fig. 3.8. Quadrupole mass filter.

positive ions tend to strike the 'positive' rods and be lost. On the y axis, the behaviour is more difficult to visualise, but in fact while the direct field tends to drive the ions out, the r.-f. field tends to focus them in. The focusing is more effective the lighter the ion, so that on this axis the heavy ions tend to be lost. Thus it can be arranged that the two conditions of stability stand, like Scylla and Charybdis (Homer), with but a narrow gap between, through which only ions of the desired type can have some chance of passing. This is indeed a filtering action, as the unwanted ions do not traverse the whole length of the device.

While small versions may easily attain a resolution up to 50,

3.5. Mass spectrometers employing radio-frequency fields

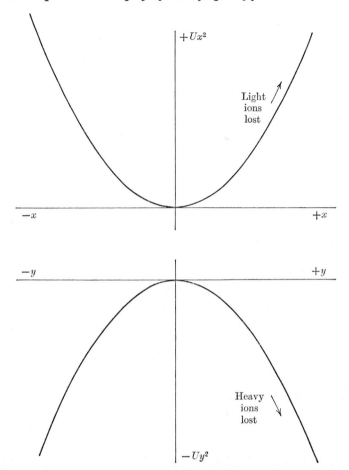

Fig. 3.9. Mean potentials in a quadrupole mass filter.

values of up to 500 are not unusual. One of the most remarkable, and often very useful features of the quadrupole mass filter is the simple control of resolution from the maximum to very low values, simply by altering the ratio V/U. The scanning of masses may be done by varying either the radio frequency, or (more often) the values of V and U, keeping the ratio at the desired value.

3.6. An unusual method of mass analysis.

Included in this summary for interest, and with a prize for simplicity, is the ion probe used on the Ariel I satellite. This consisted simply of a 9 cm

diameter sphere, extended from the satellite and biassed between
−3V and +20 V (one sweep per minute) with respect to it. Measurements of ion current were telemetered to earth. The sphere was
surrounded by a fine spherical grid (10 cm diameter) held at −6V, to
prevent excessive collection of electrons on the probe. This was the
entire 'vacuum apparatus' needed for testing the ion composition
at altitudes of about 1000 km. above the earth (see p. 44). It
is simply a variable potential barrier which is made spherically
symmetrical to be nearly omnidirectional. In this case the *velocity*

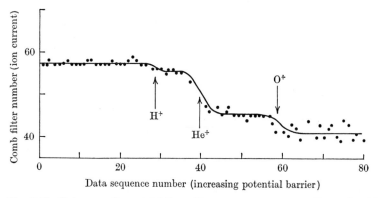

Fig. 3.10. Data scan from Ariel I satellite mass spectrometer. Example of ion
data showing presence of H+, He+ and O+ together. [From P. J. Bowen et al.
(1964). *Proc. Roy. Soc.* A **281**, 504.]

of all ions relative to the probe is closely the same, being that of the
orbital motion of the satellite (18,000 m.p.h.). Thus ions of different
mass are seen as having different energies, and may be distinguished
in a scan of retarding potential, such as that shown in fig. 3.10.

3.7. Detectors of ions. This summary would not be complete
without a brief note on the devices used for detection of ion beams
after analysis. Three types are in use.

The cup collector. The simple Faraday cup is often used, connected
to a sensitive electrometer which in the most elaborate versions
can measure down to 10^{-15} A, a current corresponding to the arrival
of about 6000 charges/s. This arrangement is simple to calibrate
and equally good for positive or negative ions. However, for the
highest sensitivities the response time is usually rather long.

3.7. Detectors of ions

The ion-electron multiplier. As indicated in fig. 3.11, a positive ion to be detected strikes the first 'dynode' of the multiplier with an energy of 5,000–10,000 eV, thereby releasing two or three (on average) secondary electrons. These are guided by the shaped dynodes to strike the next dynode in the chain, releasing on impact several more secondary electrons (with electrons, only about 200 V energy is needed for this). So the multiplication proceeds at each stage, giving at the final electrode as many as 10^6 electrons for one

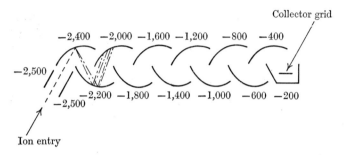

Fig. 3·11. Twelve-stage ion–electron multiplier. Typical voltages applied to each stage are indicated.

ion entering, i.e. a charge multiplication of 10^6. By this means the arrival of individual ions may be detected, and ion currents may be given as a 'count rate'. The lower limit (set by dark current and noise) probably corresponds to about 10^{-19} A, and the response time, of course, is always very short. This type of detector is always used where short response time is important, and on such instruments as the time-of-flight, no other type is possible.

The operation of such a multiplier to detect negative ions presents some practical difficulties, as a little consideration will show, but these are not insoluble.

Since the development of this, the original type of multiplier, numerous variations and improvements of design have been employed. In one of the more distinctly different variations (fig. 3.12), the ions are accelerated to energies of 30–40 keV, releasing from an electrode secondary electrons which in turn strike a phosphorescent screen with similar energy. The rate of arrival of single ions is then readily counted through the use of a common photo-multiplier to convert the scintillations to electrical pulses. Such a system is notably free of dependence on the mass and type of ion detected.

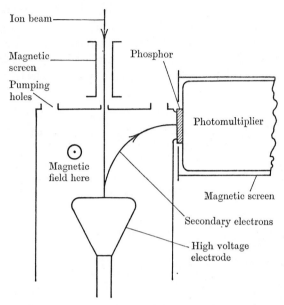

Fig. 3.12. Ion-scintillation detector. [From N. R. Daly (1965). *Nuclear Instruments and Methods*, **36**, 226; North-Holland Publishing Co.]

The photographic plate. This, the original method of detecting the arrival of ions (of suitably high energy) is still used in some spectrographs for recording a complete mass spectrum, particularly from an unsteady source.

4 The use of mass spectrometry in problems of analysis

Mass spectrometers seem to have developed such manifold uses as to rupture any attempt at rigid classification. This chapter attempts to relate some typical analytical problems to the feature uppermost in their solution by mass spectrometry, and to give some illustrative examples in most cases.

4.1. Analysis at low mass resolution. The low resolution (50 or less) achieved with some of the simplest and cheapest forms of mass spectrometer is found adequate for the two applications of leak detection and residual gas analysis. The prime requirement for both these is high sensitivity, and the ion source and mass spectrometer design are chosen with this in mind, rather than high resolution or accuracy of measurement.

Leak detection. The growth of vacuum experiments, both in number and in complexity, and in metal rather than glass apparatus, creates increasing incentives to develop methods by which any leaks in the vacuum wall may be located quickly and accurately. Small 'leak detector' mass spectrometers provide a very sensitive test, with rapid response, to such an extent that leak rates (e.g. down to 10^{-12} standard ml/s) can sometimes be measured. Such leaks are so small that it may not always be necessary to treat them as a source of vacuum failure. The usual method of operation is to connect the apparatus to pumps so that a suitable part—if not all—of the gas being pumped away passes through the mass spectrometer. This gas will consist of air if it is leaking in, but of other composition if it comes from 'outgassing' of internal components, and it should be possible to make this check initially. A fine jet of 'probe gas' is then played over the outside of the apparatus at suspect joints or parts; most commonly this gas is helium and the mass spectrometer is tuned to $m/e = 4$. The mass spectrometer output then gives immediate indication (and rough measure) of any leakage of helium into the apparatus. Hydrogen can be used as probe gas (being cheaper) but has the disadvantages of inflammability and frequently a restriction of sensitivity by the presence of a constant background peak at $m/e = 2$ (arising from water vapour and hydrogen in the residual gas). Sometimes argon is used as probe gas.

Residual gas analysis. A similar requirement of high sensitivity with only low resolution is found for the analysis of 'residual gas' (in a vacuum vessel after all the leaks have been sealed). This analysis will act both as a rough measurement of pressure in the vessel (even down to 10^{-13} torr), and possibly as an indication (by inference) of the source of the gas which is limiting the degree of vacuum. Thus it is found that water vapour is slowly released from metal surfaces for a long time, unless the whole apparatus is heated to 150–200 °C for several hours or even days, with continuous pumping. After such treatment, the range of ultra high vacuum (10^{-8}–10^{-9}) may be entered, when the principal components of residual gas are hydrogen and carbon monoxide. These gases seem to be evolved slowly from the body of metal parts (the second especially when nickel-containing alloys are present). Severe conditions of heating (up to 450 °C) are needed to reduce this source of gas to a low level and thereby to reach the range of pressures 10^{-10}–10^{-12} torr.

An interesting investigation of 'outgassing' in a small cathode ray tube (Lineweaver) showed that 95 per cent of gas released during use of the tube was oxygen, and that this arose from bombardment of the glass by the 20 keV electron beam. A complementary experiment confirmed that the glass became less dense in its surface layer, due to the removal of oxygen.

4.2. Analysis at high mass resolution.

Accurate atomic mass measurement. The first endeavours in the direction of highest possible mass resolution were made by physicists, who sought, and have succeeded in making extremely accurate measurements of the atomic masses of all the known isotopic species. In 1929, when the existence of the oxygen isotopes was demonstrated, the basis of the physical atomic weight scale was

4.2. Analysis at high mass resolution

taken as $^{16}\text{O} = 16$. Since 1961 the standard $^{12}\text{C} = 12$ has been accepted internationally, involving little change to the old chemical scale but providing in many ways a more convenient standard.

In this work, the mass spectrometer is used in such a way as to compare the masses in such 'doublets' as

$$\text{C}^{++}, \quad \text{D}_3^+ \quad (m/e = 6), \quad \text{D}^+, \quad \text{H}_2^+ \qquad (m/e = 2)$$

$$^{12}\text{CH}_4^+, {}^{16}\text{O}^+ \quad (m/e = 16), \quad ^{12}\text{CH}_2^+, {}^{14}\text{N}^+ \qquad (m/e = 14)$$

Though the members of a doublet have the same nominal mass, in each case there is a small difference which, with a mass resolution of $\sim 10^5$, can be measured with an error of less than 10^{-6} a.m.u. (atomic mass units). Thus, from the above examples, the accurate masses of H, D, ^{16}O and ^{14}N relative to ^{12}C are established, and the series can be extended in a similar manner to cover all isotopes. The results of this are concisely expressed in the packing fraction curve, given by plotting the 'packing fraction', i.e. the

$$\frac{\text{difference from nominal (integral) mass}}{\text{nominal mass}}$$

against nominal mass. This curve provides basic information in the study of the stability of atomic nuclei.

Accurate molecular weight measurement. In recent years the activity in high resolution mass spectrometry seems to be increasing, but with a shift to the field of organic chemistry, where it is a valuable tool in research into ever larger and more complicated molecules. A mass spectrum of parent and fragment ions of an unknown compound will indicate at low resolution only nominal integral values of mass, and with large molecular weights there remain many possible interpretations of the data. It is however possible to analyse the mass spectrum so exactly as regards molecular weight that most ambiguities of empirical formula are resolved at any mass number up to about 800. The determination of unique formulae for the prominent ions of the spectrum naturally assists greatly in determination of the nature of the sample (see §**5.2**).

Table 4.1 shows some of the commonest 'mass doublets' with the mass difference involved. For example, a double peak at mass 44 might be due to N_2O^+ and CO_2^+; the mass difference is due to the doublet N_2—CO and would be 11·234 milli-mass units. Assuming that a single observed mass is measured relative to a known standard of the same nominal mass, to an accuracy of 2 in 10^6, the highest mass number (M_{max}) for which the ambiguity can be resolved is shown in the last column of the table. This perhaps errs on the optimistic side as a general estimate of performance, though to demand that, in the case of a doublet being observed, the peaks must be completely resolved before measurement is possible is a pessimistic view. The limit which would be set by this view, assuming a working resolution of 30,000, is shown in the third column of the table. It is reasonable to expect the capability of such instruments to lie somewhere between the limits given.

TABLE 4.1.

Source of doublet	ΔM(a.m.u. $\times 10^3$)	M (for complete separation with resolution 30,000)	M_{max}
H_2–D	1·548	46	774
CH–C^{13}	4·467	140	(2233)
^{13}CH–N	8·109	243	
CD–N	11·028	330	
N_2–CO	11·234	337	very large
CH_2–N	12·576	376	values
NH_2–O	23·810	714	
CH_4–O	36·386	over	
H_{12}–C	93·900	1000	

4.3. 'Batch' analysis of stable samples.

Qualitative and quantitative analysis. The analysis of stable samples, in most cases a mixture of gases, is perhaps the application springing first to mind when mass spectrometers are mentioned, and this is still the job for which most instruments are designed. The overwhelming majority are of the single-focusing magnetic sector types.

Viewed purely as an analytical tool, the simplest instrument of this class will accept gaseous samples equivalent to as little as 0·2 ml. at N.T.P. (5 μmole), and produce a written record of the mass spectrum over the desired range of masses (e.g. 12–120 a.m.u.) in a few minutes; it will resolve adjacent peaks (considering integral mass numbers) up to about 200 a.m.u. The more modern or advanced instruments can produce the same information very much

4.3. 'Batch' analysis of stable samples

faster (a decade scan in less than a second), and means are provided for the recording of a wide range of ion currents, since ion peaks of only 0·01 per cent of the intensity of the base peak may be of interest. Instead of switching to different sensitivity scales by hand, this may be done automatically, or the record may be made on a logarithmic scale, or else several records are made at once with different current sensitivities.

In the majority of work, an electron impact source (see p. 22) is operated at a fixed electron energy, often 70 eV, which is one of the standard conditions under which the mass spectrum of parent and fragment ions (the 'cracking pattern') is known (or must be found) for each pure component with which the analysis is concerned. Extensive tables of such information have now been built up (see, e.g. p. 112 and table 5.1). The various peaks of the cracking pattern are expressed as percentages of the largest peak of the spectrum, so that in addition one must know the relative sensitivity of the mass spectrometer to each possible component of the mixture. Armed with this information and the observed mass spectrum of the mixture, the analysis of its composition is deduced by making the assumption that the observations are a simple sum of the (cracking pattern × sensitivity × partial pressure) for each component. For up to four or five components, depending on their nature, this may be done by hand calculation, but access to a computer will always make the task easier.

Besides the routine analyses associated with 'process control' and purity checks, samples may frequently be of the 'unknown product of a reaction' type. These are solved (if possible) by comparison of the observed mass spectrum with the tables of cracking patterns. At later stages of such an investigation the product is known, but the mechanism is investigated by various isotopic substitutions (particularly of deuterium for hydrogen) in the reactants. The distribution of isotopes in the products as shown by the mass spectrometer, is often (though not always) informative. For example, in the photolysis of cyclopentanone, ethylene is found in the product. With deuterium substitution as shown, the ex-

$$D_2C\overset{\overset{O}{\|}}{\underset{|}{C}}CD_2 \longrightarrow 2CD_2CH_2 + CO$$
$$H_2C\text{——}CH_2$$

clusive production of CD_2CH_2 argues *against* re-formation of the ring after removal of CO and before the final breakdown to ethylene. If formation of the cyclobutane ring did occur, a random distribution of isotopes among C_2H_4, $C_2D_2H_2$ and CD_4 would have been expected. Such a distribution is found in the *thermal* decomposition of cyclopentanone (Srinivasan). Generally, a non-random distribution of an isotope provides more significant information than a random one, if possibilities of other exchange reactions exist.

This may be illustrated by an experiment in which oxygen atoms were allowed to react with NO_2 (Clyne and Thrush). When ^{18}O atoms were used the reaction was stated to be

$$^{18}O + N^{16}O_2 \rightarrow 2/3(^{18}O^{16}O + N^{16}O) + 1/3(^{16}O_2 + N^{18}O)$$

The analysis of products was performed in an analytical mass-spectrometer and showed a random distribution of the oxygen isotopes. The conclusion was drawn that the reaction must proceed via an intermediate (NO_3) of D_{3h} or C_{3v} (i.e. trigonal) symmetry. However, this readily credible conclusion is only valid if one can be quite certain that there were no exchanges of isotopes in further reactions with oxygen atoms, or on surfaces during sampling and analysis.

Isotope ratio analysis. Whereas the analyses discussed above will rarely attempt an accuracy better than 1 per cent in the measurements, investigations concerned with the ratio of abundances of two isotopes of the same element may require measurement of the ratio of beam currents to ± 0.1 per cent. The earliest interests in this were, coupled with the accurate measurement of atomic masses, aimed at the determination of very accurate chemical atomic weights for the elements in their natural isotopic state. In recent years, studies of isotopic distribution in elements has become an important facet of the subject of atomic fission reactions.

A more general application of isotopic abundance measurements lies in the 'tracer dilution' technique of quantitative analysis. This has been used with radio-active isotopes, the final analysis being done with radiometric equipment. The availability of enriched or pure stable isotopes of many elements now makes wider application possible, when coupled with analysis by mass

4.3. 'Batch' analysis of stable samples

spectrometry (Webster). Briefly, the technique requires a known weight of sample in which the element sought occurs with known, usually natural, isotopic composition. This is thoroughly mixed with a known weight of tracer of the element, which has a known or measured isotopic composition considerably different from that of the sample. It is then necessary to separate some (not necessarily all) of the element from the mixture in some suitable form, and determine the isotopic composition of the result. The total weight of the element in the original sample can then be deduced.

An illustrative (and hypothetical) example may be given as follows. An analysis of a sample of aluminium alloy by spark-source mass-spectrometry (see p. 31) showed a very low proportion of copper (about 0·01 per cent or 100 ppm.) for which the empirical calibration was not as accurate as desired. The other elements found (Si, Fe, Zn) did not interfere with the copper isotopes at masses 63 and 65 a.m.u. A minute weighed sample of copper artificially enriched in isotope 65 was added to a weighed quantity of the alloy, which was then carefully remelted and further samples analysed as before. The results might be set out as follows:

TABLE 4.2. *An example of tracer dilution analysis*

	Natural Cu	Artificial Cu	Mixture
Weight of Cu	x mg in 4·972 g alloy	0·327 mg	$(0·327+x)$ mg
Measured 63/65 ratio	2·235 (natural abundance)	0·23(5)	0·88(3)

The result $x = 0·411$ mg. is readily derived, giving the proportion of copper in the original alloy as 83 ppm. Spark source analysis is not the most accurate of techniques, but a precision of ± 2 per cent in the two ratio measurements might be obtained, leading to a result correct to within 5 per cent. This would be a considerable improvement on results obtained by comparisons, in separate exposures, with samples of supposedly known composition.

Many accurate measurements of isotopic ratios now made are stimulated by geological and botanical interests. The so-called 'potassium-argon' method of determination of the age of rocks is essentially a measure of the radioactive decay of ^{40}K to ^{40}Ar in suitable minerals (mica). The ^{40}Ar thus trapped in the lattice is released by heating, and a quantitative measure of this minute sample is obtained using the mass-spectrometer. This is compared with the potassium assay of the mineral. However, an important correction must be applied for the possible presence of atmospheric argon, and the contribution of this to ^{40}Ar is represented by ^{36}Ar in known abundance ratio. A careful measurement of the ^{36}Ar/^{40}Ar ratio completes the required information.

In an elegant variation of the method (Crasty & Mitchell), the mica sample is sealed in a tube and irradiated with slow neutrons in an atomic pile. This results in conversion of some ^{39}K to ^{39}Ar, so that measurement of the proportion of the latter relative to ^{36}Ar and ^{40}Ar takes the place of the potassium assay. Samples of known age are used to provide calibration.

In another type of investigation (Urey, 1947; Epstein *et al.* 1953) minute fossil shells of known age, on a geological scale, are selected and treated to release CO_2. Extremely careful measurement of isotopic ratios, relative to a standard sample, yields a value (to ± 0.01 per cent) for the ^{16}O/^{18}O ratio in the original shells. From this, deductions may be made about the temperature of the sea (to $\pm 0.5\ °C$) at the time the fossil shell was formed. The possibility of doing this arises from the isotopic fractionation in the equilibrium of carbon dioxide and water.

$$H_2{}^{18}O(l) + \tfrac{1}{2}C^{16}O_2 \rightleftharpoons H_2{}^{16}O(l) + \tfrac{1}{2}C^{18}O_2 \quad (K = 1 \cdot 0407\ \text{at}\ 25\ °C)$$

which has a measurable dependence on temperature. The determination of 'fossil temperatures' at a variety of times and places has obvious interest.

Extensive measurements have also been made of the ^{34}S/^{32}S isotope ratios for sulphides and sulphates in the sludge from deep lakes and oil wells. The variations in this case are quite large (up to 5 per cent) with consistently high values for sulphates and low values for sulphides. The ratio for sulphur from meteorites, taken as an 'external standard' (*sic*) was intermediate. The variation has been attributed to isotope separation during bacterial reduction of sulphates to sulphides (see C. A. McDowell).

Along with these topics should be mentioned the general field of study of isotopic effects in chemical equilibria and kinetics.

4.3. 'Batch' analysis of stable samples

Differences of isotopic behaviour in chemical reactions do exist although they are generally considered small enough to be ignored. Their accurate study calls for very careful, even meticulous work; the fundamental reasons are to be found in such factors as zero-point energy of molecules, and variations of rotational and vibrational constants. In kinetic isotope effects, the parameters of the activated complex of the reaction are involved. The differences of behaviour have been used in the enrichment and separation of isotopes for use in further studies; e.g. the repeated attainment of the equilibrium

$$^{13}CO_2 + H^{12}CO_3^-(aq) \rightleftharpoons {}^{12}CO_2 + H^{13}CO_3^-(aq) \; (K = 1{\cdot}012)$$

in a counter-current apparatus, with CO_2 at 50 atm, produces an enrichment of ^{13}C in the bicarbonate phase. At the same time, ^{18}O will be enriched in the gaseous phase. The electrolysis of water—the original process for obtaining deuterium—is an example of kinetic isotope effect.

4.4. Analysis at low electron energy.

Determination of dissociation energies. The topic of this section was approached in §2.2, p. 23, in which it was seen (fig. 2.2) that the ion current for any peak decreases steeply as the electron energy is lowered to values around 10–20 eV. The 'parent ion' curves arise from a threshold (sometimes somewhat indistinct) which gives the ionisation potential (I.P.) of the molecule (within $\pm 0{\cdot}2$ eV). The 'fragment ion' curves have a threshold (usually even less distinct) which is related to the appearance potential (A.P.) of that ion from the molecule; the A.P. may usually be found to within $\pm 0{\cdot}4$ eV.

Apart from the measurement of such values, there is occasionally some advantage in the use of low electron energy to simplify a complicated 'cracking pattern' and make parent ion peaks more obvious, but the ion currents tend to be quite low and the demands on stability of the apparatus for quantitative work are increased.

The measurement of ionisation potentials of a wide range of substances is of obvious interest to those seeking to calculate these quantities by various molecular orbital methods. In the case of self-consistent field calculations, one may logically seek to compare the observed ionisation potential or potentials with the energies calculated for the electrons in each molecular orbital of the neutral molecule. An interesting set of figures for water may be quoted as an illustration (Krauss). The differences between orbital energies follow closely the differences between observed first and inner ionisation potentials, though there is an overall discrepancy of about 1·5 eV which is probably due to the neglect of correlation of the electrons in these calculations. The figures in this case can be improved to give agreement within about 0·3 eV by including estimates of the correlation energy and by making separate calculations of the ground state energies of H_2O and H_2O^+.

The procedure of making separate calculations of the orbital energies of the ion and the neutral molecule may be followed with other methods of calculation, but since the final value for ionisation potential is the difference between two relatively large numbers, errors in the calculation appear magnified. Thus comparison with experimental results is a very sensitive test of the accuracy of the theoretical methods. As explained later in this chapter, it is possible also to measure ionisation potentials of radical species by the same technique, and these results are also eagerly compared with calculations.

TABLE 4.3. *Ionisation potentials of the water molecule*

Orbital symmetry	Orbital energy (eV)	Experimental I.P. (eV)	Experimental method
$1b_1$	13·9	12·59	photoionisation
$3a_1$	15·9	14·2 }	{ photo-emission
$1b_2$	19·6	18·02 }	{ spectrometry

The deduction of thermochemical data from the measurement of I.P. and A.P. is an important topic. Two general methods of determining bond dissociation energies seem to be available, the first requiring measurement of A.P. of some fragment ion and of I.P. of the corresponding neutral radical.

For the general process

$$RX + e^-(\text{fast}) \to R^+ + X + 2e^-$$

we have

$$D(R\text{—}X) = \text{A.P.}(R^+)_{RX} - \text{I.P.}(R) - KE - E_{\text{excit}}$$

where D represents the desired dissociation energy (bond strength), KE the kinetic energy carried by the particles R^+ and X, and

4.4. Analysis at low electron energy

E_{excit} any electronic excitation. It is usually true that $E_{\text{excit}} = 0$ and that $KE < 0.2\,\text{eV}$, and failing any information to the contrary, these factors are usually ignored, though the possibilities should not be forgotten. Then if the appropriate A.P. and I.P. can be measured, the desired result follows. For example, for

$$\text{CH}_4 + e^-(\text{fast}) \to \text{CH}_3^+ + \text{H} + 2e^-,$$

electron bombardment of methane gives

$$\text{A.P. }(\text{CH}_3^+)_{\text{CH}_4} = 14.4 \pm 0.4\,\text{eV}.$$

From other experiments (see p. 92) it is found that

$$\text{I.P. }(\text{CH}_3) = 9.95 \pm 0.2\,\text{eV}.$$

Hence the result is derived

$$D(\text{CH}_3\text{—H}) = 4.45 \pm 0.5\,\text{eV} = 426\,\text{kJ/mole}$$

In cases where the ionisation potential of the relevant radical is not known, the A.P.'s of the same ion from two different reactants are measured. The difference between the two A.P. values is used, and this being easier to measure than an absolute value, the method is not grossly inaccurate. Again, other energy carried by the ions is ignored. Supporting information is needed, in the form of heats of formation of the stable compounds, which is generally available. Thus we have

$$\text{A.P. }(\text{R}^+)_{\text{RX}} = \Delta H_f(\text{R}^+) + \Delta H_f(\text{X}) - \Delta H_f(\text{RX})$$

For example $\text{C}_2\text{H}_6 + e^-(\text{fast}) \to \text{C}_2\text{H}_5^+ + \text{H} + 2e^-$;

$$\text{A.P. }(\text{C}_2\text{H}_5^+)_{\text{C}_2\text{H}_6} = 15.2\,\text{eV}.$$

$$\text{C}_3\text{H}_8 + e^-(\text{fast}) \to \text{C}_2\text{H}_5^+ + \text{CH}_3 + 2e^-;$$
$$\text{A.P. }(\text{C}_2\text{H}_5^+)_{\text{C}_3\text{H}_8} = 14.5\,\text{eV}.$$

Using the above interpretation of A.P., together with

$$D(\text{CH}_3\text{—H}) = \Delta H_f(\text{CH}_3) + \Delta H_f(\text{H}) - \Delta H_f(\text{CH}_4)$$

one arrives at

$$D(\text{CH}_3\text{—H}) = \text{A.P.}(\text{C}_2\text{H}_5^+)_{\text{C}_3\text{H}_8} - \text{A.P.}(\text{C}_2\text{H}_5^+)_{\text{C}_2\text{H}_6}$$
$$- \Delta H_f(\text{CH}_4) - \Delta H_f(\text{C}_2\text{H}_6) + \Delta H_f(\text{C}_3\text{H}_8) + 2\Delta H_f(\text{H})$$

whence the result is derived.

$$\underline{D(\mathrm{CH_3-H}) = 4{\cdot}38 \pm 0{\cdot}2\,\mathrm{eV} = 422\,\mathrm{kJ/mole}.}$$

The same information may be used to tabulate values of $\Delta H_f(\mathrm{R^+})$ for various ions; these are quite useful, and should of course be independent on the molecule giving rise to the ion (except as discussed on p. 129). Such information may then be used to test problems such as the following question of a mode of fragmentation

$$\mathrm{CH_4} + e^-_{\mathrm{fast}} \begin{array}{c} \nearrow \mathrm{CH_2^+} + \mathrm{H} + \mathrm{H} + 2e^- \\ ? \\ \searrow \mathrm{CH_2^+} + \mathrm{H_2} + 2e^- \end{array}$$

Given the observed A.P. $(\mathrm{CH_2^+})_{\mathrm{CH_4}} = 15{\cdot}3 \pm 0{\cdot}2\,\mathrm{eV}$.

$$= 1474\,\mathrm{kJ/mole}$$

and also
$$\Delta H_f(\mathrm{CH_4}) = -74{\cdot}9\,\mathrm{kJ/mole}$$

$$\Delta H_f(\mathrm{CH_2^+}) = 1390\,\mathrm{kJ/mole}$$

$$D(\mathrm{H-H}) = 426\,\mathrm{kJ/mole}$$

it is clear that at the observed A.P. there is insufficient energy for the fragmentation to produce hydrogen atoms; such a process would be expected to have an A.P. of about 19·8 eV. It is frequently possible to demonstrate the occurrence of such a second process in the mass spectrometer, and to estimate a second A.P. as shown in fig. 4.1.

Determination of electron affinity. A topic closely related to the above is the determination of electron affinities by observing *negative* ion formation under low energy electron bombardment. The electron affinity of a molecule is the exothermicity of the imagined process

$$\mathrm{X} + e^- \to \mathrm{X}^-$$

with X and X^- in their lowest electronic states. The negative ions may be produced in two distinctly different ways. The first results in formation of an 'ion pair', in

$$\mathrm{RX} + e^-(\mathrm{fast}) \to \mathrm{R}^+ + \mathrm{X}^- + e^-$$

and the variation of X^- ion current with electron energy is similar to those previously considered, the reason being that any excess

4.4. Analysis at low electron energy

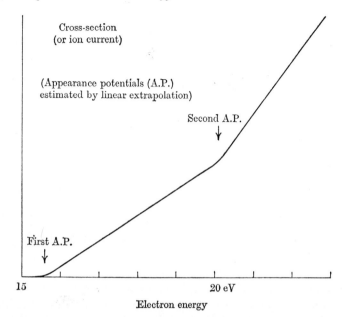

Fig. 4.1. Electron-ionisation efficiency curve for positive ions, showing two processes.

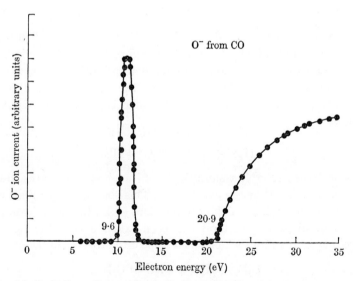

Fig. 4.2. Ionisation efficiency curve for the formation of O⁻ from CO, showing two processes. (From R. W. Kiser *Introduction to Mass Spectrometry and its Application* © 1965. Reprinted by permission of Prentice-Hall, Inc., New Jersey.)

energy of the reaction is carried away by the free electron (compare the right-hand side of fig. 4.2 with figs. 2.2 and 4.1). The other type of process is quite different, involving capture of the electron, possibly followed by dissociation.

$$RX + e^- \longrightarrow \begin{array}{c} RX^- \\ R + X^- \end{array}$$

The electron is unable to transfer appreciable kinetic energy to the heavy products, and if there is much excess energy the reaction cannot occur. Thus these 'resonance capture' reactions are only observed over a narrow range of electron energy. Fig. 4.2 shows a classic case where both types of process are clearly seen. Treatment of the data in the manner outlined for positive ions will yield values of electron affinity and the range of values is illustrated by these few examples:

TABLE 4.4. *Illustrative examples of the electron affinities of atoms*

Na, K	0·44 eV	Cl	3·69 eV
H	1·0 eV	Br	3·49 eV
O	1·45 eV	I	3·15 eV

Determination of proton affinity. Although a discussion of proton affinity does not fall strictly within the purview of this section, it seems relevant to mention here its determination by a mass-spectrometric method. This relies on the observation of ion–molecule reactions of the proton-transfer type (see p. 57), for example

$$H_3O^+ + NH_3 \rightarrow NH_4^+ + H_2O$$

or

$$H_2S^+ + H_2O \rightarrow H_3O^+ + HS$$

It seems to be a reliable assumption that if such reactions are shown to occur readily, they are exothermic. This sets an upper limit on the heat of formation of the product ion, and a lower limit on its proton affinity. The method has been taken a step further by searching for reactions which do *not* appear to proceed; for example,

$$C_2H_2^+ + H_2O \nrightarrow H_3O^+ + C_2H$$

The conclusion that such reactions must be endothermic is less soundly based, but if accepted can lead to an upper limit for proton

4.4. Analysis at low electron energy

affinity, giving from the cases quoted (Talrose and Frankevitch),

$$7{\cdot}25\,\text{eV} < P(\text{H}_2\text{O}) < 7{\cdot}44\,\text{eV}$$

or
$$P(\text{H}_2\text{O}) = 706 \pm 8\,\text{kJ/mole}$$

for the exothermicity of the imagined process

$$\text{H}_2\text{O} + \text{H}^+ \to \text{H}_3\text{O}^+.$$

In rarer cases, protonated ions may occur in normal 'cracking patterns' (e.g. H_3S^+ from thio ethers); the arguments put forward earlier may then be used directly to calculate the heat of formation and proton affinity of the ion.

The results show values of proton affinities which are somewhat surprisingly high, particularly for molecules usually thought of as saturated and stable (see table 4.5).

TABLE 4.5. *Illustrative examples of the proton affinities of atoms and molecules* (kJ/mole)

Species	Ar	N_2	H_2	CO	CH_4	H_2O	NH_3	H_2S
Limits of P	> 180	> 243	> 255 < 330	> 393	> 476 < 539	> 698 < 715	> 727	> 731 est. 840

4.5. Analysis 'on-line'

Stable samples. The term 'on-line' analysis is used to indicate the coupling of the mass spectrometer to some other system, in which changes of composition are expected to occur. The mass spectrometer is included to give immediate, continuous information on composition. For a wide range of masses, this would demand rapid repetitive mass scanning, and this can be done. However, in many cases it is sufficient to have up to three detectors set to receive simultaneously the ion currents of interest. This is done for medical purposes, for example, when an analysis of gases involved in respiration is required. By introducing a small sampling tube into the wind pipe and connecting this to a mass spectrometer a continuous record of partial pressures of carbon dioxide and oxygen is provided during the breathing cycle. An interesting example in the field of kinetics is provided by studies of the reactions of deuterium with ammonia or methane at a nickel surface (Kemball). Over periods of minutes and hours, the incorporation of deuterium into NH_3 or CH_4 could be followed, leading eventually to an equilibrium distribution among the various possible isotopic species. It was possible to trace the reaction paths to a considerable extent, and to show that they were quite different in the two cases.

A somewhat different application, now receiving much attention, is the coupling of a mass spectrometer to the output of a gas-liquid chromatography column. It has been customary to use more or less simple detectors ('katharometer' or 'flame ionisation' types) which indicate only that the effluent of the column is or is not pure carrier gas. By suitable calibration the response can be interpreted quantitatively for any compound, but the identification rests on a comparison of retention (or elution) times for known and unknown species. Particularly for complex mixtures it is possible that the retention time is not a unique characteristic for each compound, and that any one column may fail to separate all components of a mixture. The analysis of a duplicate sample on a different type of column may sometimes show the true number of components of a mixture, but still not give much extra assistance in their identification. On-line analysis by mass spectrometer, however, shows as quickly as possible the composition of each fraction of the sample as it emerges from the column. A very rapid scan through the mass range must be provided, and this is either automatically or manually initiated as each chromatographic fraction is detected by a simpler type of indicator. The mass spectrometer systems have been automated to the degree of printing out a mass spectrum at each recognisable chromatographic 'peak'.

The exact method by which the mass spectrometer is coupled to the source of gas in the above applications merits careful consideration. The gas may be allowed to leak into the ion source through a porous plug, which can be made of fine enough structure to give the 'molecular flow' conditions described in §**2.2**, p. 22, even up to sample pressures of 1 atm. In practice, a more common sampling method seems to be the 'viscous leak' in which the gas is sampled through a suitable length of fine capillary tubing (e.g. 60 cm of stainless steel tube of 0·05 mm bore). In this case, viscous flow of gas into and along the capillary should be the controlling factor, and all components of a mixture should thereby be sampled at equal rates with no disturbance of their proportions. This situation is in contrast to that which obtains when molecular flow occurs,

4.5. Analysis 'on-line'

where (since molecular velocities are involved) the sampling rate is biassed in favour of lighter components by a (molecular weight)$^{-\frac{1}{2}}$ factor. However, it is important to note that, whatever the mode of ingress, the gases are always pumped away from the ion source under conditions of molecular flow. This gives the result that the *observed* partial pressures in the source are a *correct* representation of the sample when the entry of gas is controlled by *molecular flow*, but are biased in favour of heavier components for the case of the 'viscous leak', when a correction factor of (molecular weight)$^{-\frac{1}{2}}$ must be included. It should also be noted that there are some difficulties in obtaining conditions with a simple capillary tube which correspond truly to 'viscous flow' sampling, since the flow must at some point in the tube change to 'molecular' type as the pressure falls. It is all too easy to achieve a sampling condition which is a complicated mixture of the two basic types, and the design of these sampling leaks must be done with some care to obtain the desired effect (see McDowell, § 8.10).

Unstable samples. This section is clearly divided from the previous one in that it describes methods and apparatus developed for the investigation of unstable chemical species, chiefly radicals. These methods, of course, can be used equally well for stable species. With unstable samples the most direct communication possible is needed between the system under test and the mass spectrometer—ideally a collision-free path is sought. The arrangements evolved often bear a resemblance to those used for the sampling of ionic species, one example of which is illustrated in fig. 2.11. The sample must first be formed (at least crudely) into a molecular beam and this then directed into an ion source, the ions so produced being analysed with a mass spectrometer. An important advance in this problem has come with the development of 'nozzle' sources (Kantrowicz and Grey), in which a considerable collimation of the beam is achieved (with a narrowing of the energy distribution) from the expansion of a relatively high-pressure gaseous sample through a correctly contoured nozzle. The supersonic jet so formed, in a domain of quite low pressure, is aimed at further apertures, the first of which is called a 'skimmer', which can then be made to define quite efficiently a reasonably intense beam of molecules from the original source.

It should be noted that the considerable advantages of ion lenses and focusing potentials as used in figs. 2.9 and 2.11 are of no avail here, and that the process of ionisation is usually an inefficient one, perhaps occurring with only 1 in 10^4 of radicals entering the ion source. Thus although systems such as flames and shock tubes can generate neutral radical species in concentrations greatly in excess of ionic species, the latter are probably easier to detect and analyse with a mass spectrometer. Nevertheless, some success has been registered in the investigation of radical species in these systems, and also in electric discharges, their afterglows, and in pyrolysis and photolysis experiments.

The last mentioned have been the most fruitful in producing particular radicals for study. The quantity most often sought is the ionisation potential of the radical, and the method of measurement and the uses of this quantity were mentioned on p. 84. Data on a great number of radicals have been produced, and this accumulated pattern, even more than individual results, has been of great interest to those studying the structure of molecules and radicals from theoretical aspects. Certain radicals are especially amenable to theoretical treatment, and in these cases remarkably good agreement between calculated ionisation potentials and the experimental values has been demonstrated (Hush & Pople).

An example of the apparatus used in such experiments is shown in fig. 4.3. The gas chosen for the experiment is diluted (about 1000:1) with helium 'carrier' gas, which then flows into the central tube at pressures around 10 torr. The residence time in the 'reaction zone' is about 10^{-3} s, during which the sample is either heated or strongly illuminated with a mercury lamp. In the latter cases the reaction is 'sensitised' by saturating the gas with mercury vapour. Conditions are adjusted to an optimum for production of the desired radical, a sample of which, after effusing through the fine hole (0·03 mm diameter) in a quartz thimble, passes directly into the ion source at pressure not greater than 10^{-4} torr. With such a free choice of starting material and conditions, quite large concentrations of radicals (at least 1 per cent of reactant) may be produced in the reactor. However, the experiment is not thereby necessarily made easy since, if the compound used does dissociate easily to the radical this process may as readily occur at the filament or other hot metal of the ion source or by electron bombardment,

4.5. Analysis 'on-line'

giving false results unless particular precautions are taken (Barber et al.).

It is important to realise that the very detection of free radicals almost always relies on the use of low electron energies, within a few volts of the ionisation potential. For success with this method,

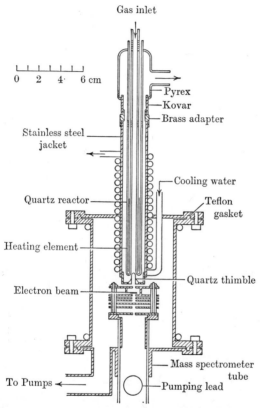

Fig. 4.3. Apparatus for the sampling of radicals from pyrolysis reactions. Arrangement of reactor and ion source. [From F. P. Lossing & A. W. Tickner (1952). J. chem. Phys. 20, 908].

there should exist some workable range of electron energies within which ions can only arise directly from the radical, and not from some fragmentation process of a heavier species, or from another ion of the same mass number. The detection of ions when electrons within this energy range are used is the only unambiguous proof of the presence of neutral radical species *in the ion source*. As mentioned above, there are still spurious modes of formation of radicals to be considered before the proof is related to the system under study. The 'available ranges' of electron energy are shown in table 4.6 for some important cases.

TABLE 4.6. *Appearance potentials relevant to the detection of radicals*

Mass number	Process	Electron energy threshold (eV)	'Diagnostic range' of electron energy (eV)
1	H^+ from H	13·6	—
	H^+ from H_2	18·1	4·5
	H^+ from H_2O	19·6	—
14	N^+ from N	14·54	—
	N^+ from N_2	24·3	9·8
15	CH_3^+ from CH_3	9·95	—
	CH_3^+ from CH_4	14·4	4·45 in methane
	$^{13}CH_2^+$ from CH_4	15·6	—
	CH_3^+ from C_2H_6	13·95	4·0 in ethane
16	O^+ from O	13·61	—
	O^+ from O_2	16·9	3·3
	O^+ from H_2O	18·9	—
	NH_2^+ from NH_3	15·8	2·2 if NH_3 present
17	OH^+ from OH	13·0	—
	OH^+ from H_2O	18·8	5·8 in water
	NH_3^+ from NH_3	10·52	impossible if NH_3 present
29	$C_2H_5^+$ from C_2H_5	8·7	—
	CHO^+ from CHO	9·88	difference 1·2
	$C_2H_5^+$ from C_2H_6	12·8	3·1 or 1·9
	$C^{13}CH_4^+$ from C_2H_4	10·5	1·8 or 0·6 with C_2H_4 present
33	HO_2^+ from HO_2	11·53	—
	$^{17}O^{16}O^+$ from O_2	12·2	0·7 with oxygen
	HO_2^+ from H_2O_2	15·4	3·9 with H_2O_2

In the 'flow reactor' experiments described above, the reactant and conditions can usually be chosen to assist this requirement of a 'diagnostic range' of electron energies. The situation is often much more difficult in the study of flames and electrical discharges as such, when the system must be accepted as it is. Largely for this reason, the study of radicals in low pressure flames has provided firm confirmation of the presence of only a few simple radical

4.5. Analysis 'on-line'

species: H, O, OH, and HO_2 in hydrogen–oxygen flames, and HO_2 CH_3 and CHO in methne–oxygaen flames. The reports of radicals CH_2O and CH_3O in the latter seem less firm, though such species and many more would be expected in such a reactive system. The table above indicates how much easier is the detection of CH_3 in the methane–oxygen system than C_2H_5 would be in ethane–oxygen mixtures (where CHO and C_2H_4 may also be produced).

The gases flowing out of an electric discharge region have been studied much more than the discharge plasma itself, as regards radicals. The species reported have again been very simple: H, O, OH, N, NH_2, N_2H_3 and HO_2. In some experiments these species have been present in large enough concentration for quantitative measurement with electrons of $\sim$ 50 eV energy, being treated as a component of the mixture, in the normal 'analytical' manner. In other instances study with low electron energies has been able to demonstrate the presence of electronically excited states of N and of N_2, through the observation of ionisation potentials which were too low to be considered due to such species in ground states. Undoubtedly much more work will be done with these flowing systems, particularly with the ideas of 'chemical manipulation' mentioned in the context of ion-molecule reactions (p. 51). The progress in the study of flames is likely to be slower, in view of their greater complexity.

The last application to be mentioned is one which might admittedly verge sometimes on the 'stable sample' class, but is included here. This topic of 'Knudsen cells mass spectrometry' has evolved in a way of its own, for the study of high temperature reactions and equilibria. It has revealed numerous species which, while stable in the vapour phase at high temperatures, are otherwise unknown or uncommon (McDowell; Reed).

For these experiments a Knudsen effusion cell, which can be heated to a (measured) temperature of up to 2500 °K, is built close to a conventional form of electron impact ion source. A typical apparatus is depicted in fig. 4.4. The object of the method is to achieve conditions of thermal equilibrium between vapour and solid inside the cell, at the known temperature. The vapour species effuse through a very small orifice, and are identified and the relative partial pressures measured by normal use of the mass spectrometer. It is important that the effusion of species should not disturb the equilibrium, and a large ratio of surface area of the sample to the area of the orifice is always used. The calibration of pressures in absolute terms can be obtained by comparison with a substance of known behaviour, or alternatively by an integration of ion current during the complete evaporation of a known weight of a standard (e.g. silver) sample from the cell.

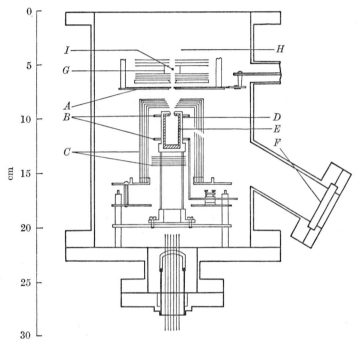

Fig. 4.4. Ion source with Knudsen effusion cell. A, movable slit, plate and mount; B, tungsten filaments; C, tantalum radiation shields; D, tantalum Knudsen cell; E, graphite liner; F, quartz window; G, ionisation chamber; H, source focusing system; I, ionising electron beam (cross-section). [From W. A. Chupka & M. G. Inghram (1955). *J. Phys. Chem.* **59**, 100.]

In one of the most important experiments of this type (Chupka & Inghram, 1955), the cell was lined with graphite and the variation of vapour pressure with temperature was studied. This led to a direct determination of the much-disputed latent heat of vaporisation of graphite (now taken as 710 kJ/mole). In the same work, it was shown that the vapour is a mixture of the species C, C_2 and C_3, with some C_5 and heavier species. The value quoted

4.5. Analysis 'on-line'

above applied to C (gas) only, and further values of 794 and 836 kJ/mole were determined for the heats of vaporisation of C_2 (g) and C_3(g) respectively. These values were obtained both from plots of log (relative partial pressure) versus $1/T$ (second-law method) and from calculations based on absolute pressure (third-law method).

In a modification of the above experiment (Berkowitz), nitrogen was passed slowly through the graphite-lined cell, and the CN radical was observed and its heat of formation measured. When hydrogen was passed over graphite, at least five different hydrocarbon species were found (CH, CH_2, CH_3, C_2H, C_2H_2).

A somewhat more difficult system was afforded by sulphur, in which all species from S_2 to S_8 were observed (Berkowitz & Marquert). The fragmentation of the heavier species during ionisation to give contributions to the lighter ones was a considerable trouble, but heats of formation and equilibrium constants for the various reactions were determined. An interesting comparison was made with 'free evaporation' of sulphur, i.e. evaporation from sulphur not completely confined in a cell. The vapour composition was then markedly different, and depended on the solid sample used. Rhombic sulphur gave largely S_8 molecules and 'Engels sulphur' gave largely S_6 molecules. These observations must reflect on the actual structure of the solid form, and underline the importance of achieving a true thermal equilibrium in the cell (Berkowitz & Chupka).

In many other experiments on metal oxides, sulphides and halides, the pure substances have been placed in the cell and the composition of the vapour studied at various temperatures; a surprising variety of species have been found to occur. For example, over BaS were found the species BaS, Ba, S, S_2 and Ba_2S_2. A most useful fund of thermochemical data in the form of heats of formation and heats of dissociation of such molecules in the vapour phase has been derived in this manner. Some work has turned attention to pure metals, and the heats of sublimation of zinc and cadmium have been determined (Mann & Tickner). Such experiments, in which equilibrium vapour pressures of metals over alloy systems are examined, might in due course give a complete picture of the variation of chemical potential of the components of an alloy through wide ranges of composition.

5 The interpretation and prediction of mass spectra

5.1. Introduction. This chapter concerns itself with the topic of the mass spectra of polyatomic molecules, the so-called 'cracking patterns', as generated by the most common types of mass spectrometer (§ 3.2). These spectra are largely a characteristic of the parent molecule but are also dependent on numerous instrumental factors. There is at present a marked dichotomy of approach to the data.

On the one hand is the empirical approach of the analyst, who can make effective use of tabulated data in analysis of gas mixtures (see p. 78). In the case of pure compounds sent for analysis, it is obvious that the cracking pattern contains much more information than simply the molecular weight of the compound. Experience and familiarity enable the recognition of features in the mass spectrum which reflect the molecular structure of the parent compound, and can often give a more or less complete determination of it. The supposed breakdown sequence of the parent into fragment ions is rationalised by consideration of likely mechanisms, and the discovery of many regularities of behaviour. This line of approach seems most fruitful when applied to molecules of moderate size (say molecular weights of 100–1,000) though the upper limit particularly is being extended.

The other approach is that of the theoretical or physical chemist, who takes up the challenge of explaining the observed mass spectrum in fullest possible quantitative detail, with the goal eventually of predicting mass spectra and accounting for instrumental factors. As will be seen, this leads into involved statistical analysis of most tortuous problems of chemical kinetics, and it must be stated at the outset that the theory is in the early stages of development. The aspirations are at present limited to the discussion of fairly simple compounds, but there is hope that a firm foundation, once laid, would bear a considerable scientific edifice.

5.1. Introduction

Since the theoretical approach introduces ideas which are helpful in the other aspect, it will be considered first.

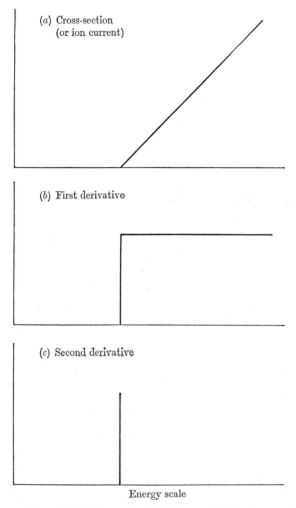

Fig. 5.1. Ideal linear threshold law and derivatives.

5.2. A second look at ionisation efficiency curves.
It seems appropriate to commence with a closer look at the probability of producing an ion in a bombardment event. In the case of electron bombardment, the 'linear threshold law' has already been illu-

strated in figs. 2.2 and 4.1 and its excellent experimental confirmation, for atomic species at least, has been mentioned (see p. 25). The ideal form of the law, for a *single* transition, is shown again in fig. 5.1 (*a*) together with its first derivative (*b*) and second derivative (*c*). Quantum mechanics has provided theoretical justification for the form of fig. 5.1 (*a*) for electron bombardment (see Art. 8 of Reed)

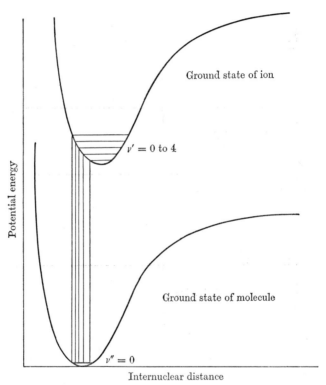

Fig. 5.2. Vertical transitions in ionisation.

and for (*b*) for the case of photon bombardment (compare fig. 2.3) for atomic and simple molecular species. However, these threshold laws are not yet fully confirmed for large molecules. Notice now that the second derivative of the electron ionisation efficiency curve, and also the first derivative of the photoionisation efficiency curve, shows just the energy of the transition responsible; these will in later (more complicated) cases be referred to as 'second derivative curves'. In the case of diatomic molecules, the possibility already

5.2. A second look at ionisation efficiency curves

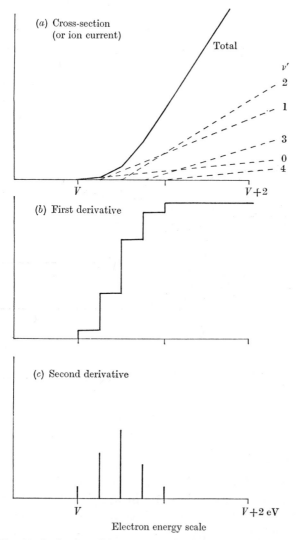

Fig. 5.3. Ionisation efficiency curves for fig. 5.2 and derivatives.

exists of several transitions to upper vibrational levels of the ion, as illustrated in fig. 5.2. The ionisation occurs so rapidly that the Franck–Condon principle should govern the relative probability of the possible transitions, in the manner familiar in optical spectroscopy. The example illustrated in fig. 5.2 is taken further in fig. 5.3 to demonstrate that the ionisation due to each transition will obey the simple law, with different relative probabilities, to yield a total ionisation efficiency curve as in fig. 5.3(a). This segmented type of curve has been obtained in experiments using monoenergetic electron beams, while the first derivative curve (b) has been observed in photoionisation experiments. Fig. 5.3 (c), the 'second derivative curve' reproduces the energies of the transitions and will also indicate their relative probabilities (Franck–

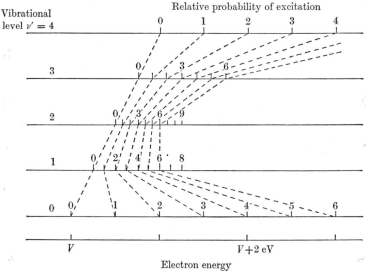

Fig. 5.4. Probability of excitation of an ion versus electron energy (following figs. 5.2 and 5.3, with vibrational quantum = 0.25 eV).

Condon factors). Although the theory suggests that the second derivative should consist of a set of delta functions, of indefinite height, the height and width are in practice always finite, since there must be some distribution of electron energy which is revealed in the second derivative.

One further type of plot is derived from this example, as a useful illustration. Fig. 5.4 is drawn to show, on lines representing the energy of each vibrational level of the ion, the relative probability of excitation to that level as a function of electron energy. Points of equal probability are joined by dotted 'contour' lines. This is done to indicate the construction of a more complicated 'energy transfer' diagram to be discussed in the next section.

5.3. A second look at ionisation efficiency curves

Changes of rotational state are not strongly induced by electron bombardment, and effects due to transitions of this type are hardly observed; changes of translational motion are likewise negligible.

5.3. Internal energy in polyatomic parent ions. For the polyatomic molecules which are the goal of this discussion, the ionised state of the molecule will possess a very wide selection of more or less closely spaced vibrational levels, and may even multiply these by having several low-lying electronic states of the ion. It becomes impossible in such cases to discuss individual transitions, and the simplest approximation is to assume a continuum of levels within a certain energy band or bands. The formation of parent ions with internal energies in excess of that of the ground state of the ion may then be discussed as before.

Following the arguments of the previous section, one may suppose that careful measurements of ionisation efficiency are made on some molecule, by a method using monoenergetic electrons. Although fragmentation may occur, it is the total ionisation that must be measured for this purpose. The second derivative of this will then give a picture of the band structure for the absorption of energy in ionising collisions with electrons, which will be referred to as an 'absorption band diagram'. An example of this is given in fig. 5.5(a) which is a slightly smoothed version of experimental results on propane, this being a molecule much discussed in the context of this theory. From this is derived the 'energy transfer diagram' [fig. 5.5(b)] which is analogous to fig. 5.4. The importance of this is in showing, for any selected electron energy, the distribution of values of total internal energy (E_i) in the parent ion (assuming the linear threshold law to hold for every transition). As will be discussed, it is this parameter more than any other which controls the rate of each mode of fragmentation—one example of a corresponding variation of dissociation rate constant is shown on the right-hand ordinate.

5.4. Experimental information on excited states of ions. A brief note is interpolated here on the experimental investigation of excited states of molecular ions. The method mentioned above, involving the second derivative of an experimental curve, imposes severe requirements on the stability and general design of the

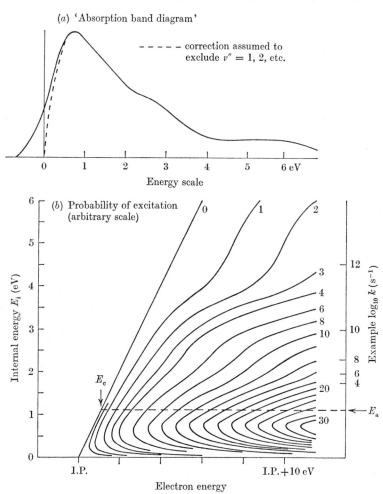

Fig. 5.5. (a) Experimental (total) second derivative curve for propane.
(b) Derived energy-transfer diagram.

apparatus if the results are to be reproducible and reliable. An attractive alternative is to study photoionisation, since the first derivative then gives the type of information required. Since, however, the transition probabilities may differ markedly between the electron and the optical process, the results in this respect may truly be different.

The form of the second derivative curves is often in satisfactory

5.4. Experimental information on excited states of ions

accord with calculations of electronic energy levels for the ion, based on the molecular orbitals of the neutral molecule. The discrete levels so indicated are widened into energy bands by the various possibilities of vibrational transitions, which are weighted both by the relevant Franck–Condon factors and by the Boltzmann distribution among the levels of the neutral molecule. The sum of several bands leads to a curve such as that of fig. 5.5(a), as fig. 5.10 shows.

Quite recently, a new method for the investigation of excited states has emerged, under the names of photo-electron spectroscopy, or photo-emission spectrometry, which promises to be a very powerful tool in this quest. Briefly, the gas under study, at low pressure, is irradiated with a single well-defined emission line from a discharge light source (usually He at 584 Å). Photoionisation of the target molecule may occur to the ground state of the ion, or to various electronically excited states. The occurrence of photoionisation is accompanied by emission of an electron with kinetic energy equal to the difference between the energy of the photon (which is accurately known; 584 Å $\equiv$ 21·21 eV) and the ionisation energy of the ion. Thus careful measurement of the energy spectrum of the electron photo-emission current gives the required information on the excited states of the ion. Examples of such energy analyses are shown in fig. 5.6. The first example shows the current of electrons able to cross a potential barrier which is varied and set to known levels. From the curves shown for PH_3 and NH_3 the energies required to form the ground and first excited states of the ions PH_3^+ and NH_3^+ may be inferred, with an accuracy of about ± 0.05 eV. In fig. 5.6(b) is shown a curve taken with hydrogen in another apparatus, from which the energies of the vibrational states of H_2^+ may be measured. The different (in fact, differential) form of curve arises from the use of a deflection energy analyser (see §3.3) and achieves an energy 'resolution' of 0·1 eV. Developments may be forseen leading possibly to resolution as precise as 10^{-4} eV, which should reveal rotational structure. The height of the 'lines' in fig. 5.6(b), or the steps in 5.6(a), give the relative Franck–Condon factors for the transitions involved.

The one apparent lack in this very powerful new tool seemed to be a check on the nature of the ion produced. This need appears to have been most elegantly met in some recent experiments

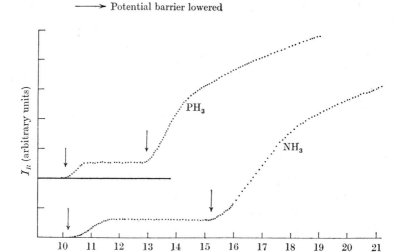

Fig. 5.6(a). Photoelectron stopping curves for ammonia and phosphine. [From D. C. Frost et al. (1968). *Adv. in Mass Spectrometry* **4**, Inst. Pet.]

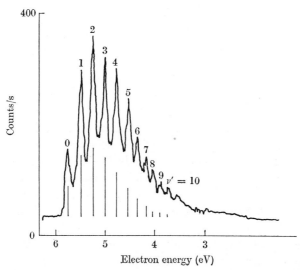

Fig. 5.6(b). Photoelectron spectrum of hydrogen excited by helium resonance radiation ($h\nu = 21{\cdot}21$ eV) The vertical lines represent calculated values of the Franck–Condon factors (see text). H_2^+, $X^2\Sigma_g^+ \to H_2$, $X^2\Sigma_g^+(\nu'' = 0)$. [From D. W. Turner & D. P. May (1966). *J. chem. Phys.* **45**, 473.]

5.4. Experimental information on excited states of ions

(Brehm & von Puttkamer) in which 584 Å radiation is focused into a somewhat modified ion source of a mass-spectrometer. The photoelectrons travel out of the source in the opposite direction to that taken by the positive ions, and meet a variable potential barrier which forms an energy analyser. Any electron which can cross

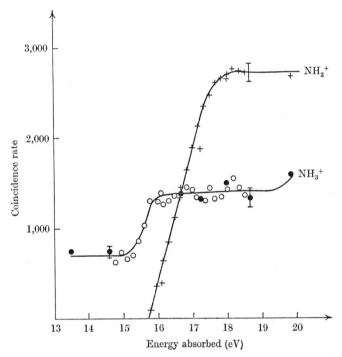

Fig. 5.7. Photoelectron stopping curves for specific ions from ammonia, with 584 Å radiation. [From B. Brehm & E. von Puttkamer (1968). *Adv. in Mass Spectrometry* **4**, Inst. Pet.]

this barrier registers a 'count' through a multiplier detector. At the same time, the mass spectrometer is operating in the normal way, and ions of a chosen m/e ratio can register 'counts' at another detector. The final output shows the rate at which coincidences of such events occur, and the result appears as in fig. 5.7. The value of the additional information is evident when this is compared with fig. 5.6(a), showing clearly the occurrence of the process.

$$NH_3 + h\nu \rightarrow NH_2^+ + H + e^-$$

at energies of 15·7 eV or greater, which was not evident in the less elaborate experiment.

5.5. The Quasi-equilibrium theory. (see McLafferty; Reed). The core of the present theory of mass spectra is the step which relates the internal energy of the parent ion to its rate of decom-

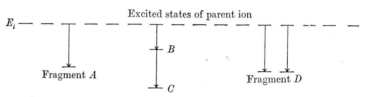

Fig. 5.8(a) Diagram of the decomposition of some states at internal energy E_1

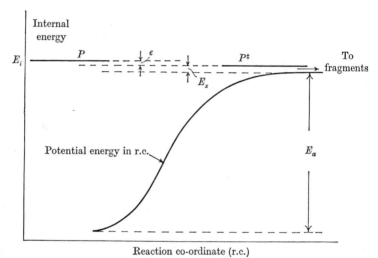

Fig 5.8(b) Subdivision of the internal energy of an ion. P shows the internal energy of non-active states; $P^{\ddagger}$ the internal energy of an activated state less the translation energy in the reaction coordinate.

position in some specified way. It is assumed that at any given internal energy (E_i), many states of the ion can exist, each with the energy variously distributed. In most cases considered there are several modes of decomposition proceeding simultaneously, each with its own dissociation rate, to give different fragment ions. Further, the lifetime of the parent ion is assumed to be much greater than any of the vibrational periods, so that a statistical

5.5. The Quasi-equilibrium theory

distribution of ions among the states is attained (only thus is the internal energy the only parameter needed to describe the ion). The situation is indicated diagrammatically in fig. 5.8(a).

It is reasonable to suppose that only a proportion of the many states will correspond to 'activated complexes', able to decompose to one fragment or another. The key assumption is now made that commutation between the states is more rapid than any decomposition, so that the population of an activated state is not appreciably depleted by its irreversible conversion to fragment ions. This gives the 'quasi-equilibrium' among the states of energy E_i from which the theory takes its name and failing which the probability of decomposition might be a complicated function of time, not able to be expressed as a unique and simple rate constant.

With the above model of the situation in mind, the decomposition of an activated state is tractable by a slight modification of absolute reaction rate theory. If the height of the energy barrier to decomposition is taken as E_a, then the energy remaining in the degrees of freedom of the activated complex is $(E_i - E_a)$. This may be further divided so that the ('excess') energy remaining in the internal degrees of freedom (rotation and vibration) of the complex is called E_x, while the reaction coordinate (best thought of as the length of the bond which is breaking) is considered to carry a translational energy ϵ as the top of the energy barrier is reached. Thus it is true that

$$\epsilon = E_i - E_a - E_x \tag{5.1}$$

The relationships of the symbols are further explained in fig. 5.8(b).

The contribution to the overall dissociation rate through a *single* activated state may first be considered, i.e. one allowed value of E_x is selected. Then for molecules having internal energy between E_i and $E_i + \delta E$ (where necessarily $E_i \geqslant E_a + E_x$), the values of ϵ will also lie in a band δE wide. The step now taken is to consider a portion of the reaction path of length 'l' at the top of the barrier where the potential energy varies very little with distance. These conditions are compared with those of translational energy quantised in a box of definite length 'l'. For a particle of mass m in a box, the well-known formula for translational energy

$$\epsilon = \frac{n^2 h^2}{8ml^2}$$

may be transposed to give the quantum number of any level as

$$n = 2l/h \sqrt{(2m\epsilon)}$$

The energy difference between successive levels, which may be interpreted as the density of these levels $\rho(\epsilon)$ is then

$$\frac{dn}{d\epsilon} = \rho(\epsilon) = \frac{2l}{h}\sqrt{\frac{m}{2}} \cdot \epsilon^{-\frac{1}{2}} \quad (5.2)$$

(This may be considered to mean that higher values of ϵ are less probable than lower ones.) The contribution to the overall dissociation rate (r) at an energy ϵ is taken as the reciprocal of the 'time to cross the box'

$$r(\epsilon) = \frac{1}{l}\sqrt{\frac{2\epsilon}{m}} \quad (5.3)$$

It thus appears that the increased rate of crossing at high energies exactly counter balances the lower probability of these energies, giving for the total dissociation rate in a band δE wide

$$r = \tfrac{1}{2}\rho(\epsilon) \times r(\epsilon)\,\delta E = \frac{\delta E}{h} \quad (5.4)$$

where a factor $\tfrac{1}{2}$ is included so that values of ϵ in only one direction are considered.

Now the specific dissociation rate k_j through the selected activated state (j) must refer to the whole ensemble of states and contain a factor expressing the rarity of the state j, i.e.

$$k_j = \frac{1}{h} \times \frac{1}{\rho(E_i)}$$

where $\rho(E_i)$ is the density of all states of the ion at an internal energy E_i. The result is easily generalised to consider the total number of possible activated states $N(E_i - E_a)$ having E_x between zero and $(E_i - E_a)$.

$$k = \frac{1}{h} \times \frac{N(E_i - E_a)}{\rho(E_i)} \quad (5.5)$$

If N is sufficiently large this may be written as

$$k = \frac{1}{h\rho(E_i)} \int_0^{E_i - E_a} \rho^{\ddagger}(E_x)\, dE_x \quad (5.6)$$

where $\rho^{\ddagger}$ is the density of states in the coordinate E_x. However,

5.5. The Quasi-equilibrium theory

in the regions of energy which are of most interest, the use of this 'integral approximation' is highly questionable.

Such expressions for k are of little use unless values for the density of total states $\rho(E_i)$ and density of states of activated complex $\rho^{\ddagger}(E_x)$ or their number $N(E_i - E_a)$ can be inserted. The difficulties in even estimating these functions for the ion and its activated complex, when accurate values of the internal rotational and vibrational parameters are just not available, seem almost insuperable. One simplification usually adopted is to ignore the factors allowing for many electronic states, as these should cancel out unless stable states of the ion include access to several electronic states, while the activated complex is confined to one configuration.

For the purposes of illustration, it does not seem to the author helpful to quote more than the simplest formula obtained by applying (5.6) to the dissociation of a parent ion having s simple harmonic oscillators, which is

$$k = \nu \left(\frac{E_i - E_a}{E_i} \right)^{s-1}$$

The derivation of this uses the integral approximation, and ignores contributions from rotational degrees of freedom. As a result, the number of 'effective' oscillators of the molecule (s) and the frequency factor (ν) must be treated as parameters, and adjusted to a tolerable fit with the data. More rigorous but much more involved formulae have been derived which consider internal rotational motion and a true summation of the available states.

A numerical example may be quoted for the case of a decomposition of the propane ion:

$$C_3H_8^+ \rightarrow C_2H_5^+ + CH_3$$

The estimated form of the rate constant formula, as used for the calculations of fig. 5.9(b) was

$$k = 1 \cdot 12 \times 10^{15} \left(\frac{E_i - 1 \cdot 1 \text{ eV}}{E_i} \right)^{25}$$

from which also the example displayed in fig. 5.5(b) was calculated.

5.6. The comparison of experiment with calculations.

The treatment of experimental data. The simplest presentation of experimental data is the 'cracking pattern' for a specified electron

TABLE 5.1. *Cracking pattern (70 eV) and appearance potentials of ions from* n-*propane*

Mass number	Species	Relative abundance		AP. or I.P.
		Observed	Calculated	
44	$C_3H_8^+$	12·7	10·3	11·07
43	$C_3H_7^+$	10·6	14·4	11·7
42	$C_3H_6^+$	1·5	1·9	12·2
41–36	$C_3H_5^+$—C_3^+	12·5 total	9·8 total	
29	$C_2H_5^+$	31·1	25·7	12·2
28	$C_2H_4^+$	19·2	22·3	11·7
27	$C_2H_3^+$	10·5	12·8	15·3
26	$C_2H_2^+$	1·9	2·8	14·5

energy. This is shown in table 5.1 for the much discussed case of propane, at an electron energy of 70 eV.

A much more detailed comparison with theory is possible if the variations of ion currents near the threshold of ionisation (10–20 eV) are considered. This is also more satisfactory from the point of view of the theory, which is much more likely to be applicable in this region, where the linear threshold law is at least approximately true. The experimental data are most easily compared when presented as a 'second derivative curve' for parent and each fragment ion, on a scale of electron energy.

This is usually presented in a normalised form, and called a 'breakdown curve'; fig. 5.9(a) shows the example of this for propane. It is interesting to see also the un-normalised data (fig. 5.10) which of course fit into the 'total' envelope used in fig. 5.5(a). If the latter is considered to portray the pattern of energy states in the excited propane ion, then fig. 5.10 suggests the predominant products of each energy absorption band.

As mentioned previously, such curves have emerged from careful measurements both with mono-energetic electron impact ionisation and with photoionisation. A further method has been the use of charge-exchange and proton-transfer reactions (see p. 56 and 57). In these experiments, falling into the class considered in §2.4, p. 52, ions of a chosen type are passed into the gas under study, and by electron or proton transfer generate an ion whose excess energy E_i is more or less accurately known to be the difference between the I.P. of the new ion and that of the original ion

5.6. Comparison of experiment with calculations

(assuming that both ions are in ground electronic states). The extent of decomposition of the new ion is observed, and the experiment repeated with as many other bombarding ions as possible.

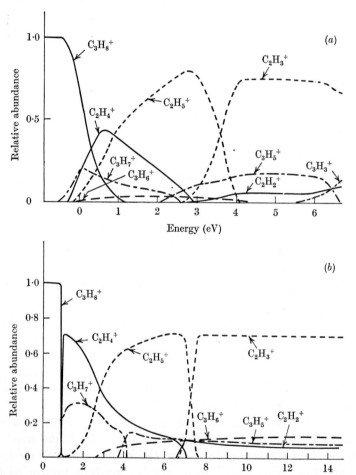

Fig. 5.9. Propane breakdown curves: (a) experimental, (b) original calculation. [From H. M. Rosenstock (1963). *Mass Spectrometry of Organic Ions* (ed. F. W. McLafferty) Academic Press, p. 30.]

Each experiment gives a set of points on an experimental breakdown curve, plotted against excess energy E_i, which is quite similar to the others discussed above (Wilmenius & Lindholm; Perrerson & Lindholm).

Predictions from the Quasi-Equilibrium theory. The further application of the quasi-equilibrium theory requires first the selection of a detailed 'breakdown scheme' for the parent ion. This must be designed to represent the main reaction paths leading to each ion, but cannot seek to include all possible paths for reasons of complexity. There are frequently ambiguities which make the selection of a scheme somewhat arbitrary, but for propane the following seems agreed, and provides a good example.

$$C_3H_8^+ \to C_3H_7^+ \to C_3\dot{H}_5^+ \to C_3H_3^+ \to C_3H^+$$
$$\hookrightarrow C_3H_6^+ \to C_3H_4^+ \to C_3H_2^+$$
$$\hookrightarrow C_2H_5^+ \to C_2H_3^+ \to C_2H^+$$
$$\hookrightarrow C_2H_4^+ \to C_2H_2^+$$

The most direct confirmation of some processes comes from the study of metastable decompositions (see §**5.7**).

Now, for each of the steps involved in the scheme, the best possible calculations of the rate constants as a function of E_i must be made, leading to an 'elementary' mass spectrum for that value of E_i; the calculations are then repeated for other values of E_i throughout the region of interest. The direct plot of these figures against E_i should yield the 'breakdown curve', and an example of such a calculation is shown in fig. 5.9(*b*) for comparison with the experimental result. As will be realised from the next section (§**5.7**) it is necessary in these calculations to assume a definite value for reaction time if the rate constants are of low values, but otherwise only a ratio of rate constants is required.

For the calculation of the mass spectrum at any electron energy (e.g. 70 eV) the energy transfer diagram [as in fig. 5.5(*b*)] must be brought into the calculation. For the chosen value of electron energy, the 'elementary' mass spectra must be multiplied by the weighting factor appropriate to their value of E_i, as derived from the diagram. A summation is then carried out over all relevant values of E_i. The result of such a calculation is shown in table 5.1.

It is interesting to consider that the energy transfer diagram displays as nearly as is possible a variation of 'temperature' of the parent ions with electron energy. Since the ions should

5.6. Comparison of experiment with calculations

follow a collision-free path through the mass spectrometer, each is an isolated system, and temperature in the normal sense has no meaning for the decomposition of the ions. It is true that ion source

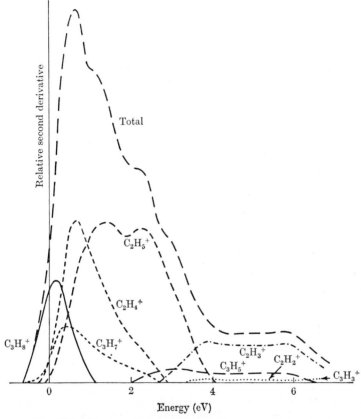

Fig. 5.10. Un-normalized second derivative curves for propane. [After W. A. Chupka and M. Kaminsky (1961). *J. chem. Phys.* **35**, 1991.]

temperature has an effect on fragmentation, but this may be seen now as an indirect effect, traceable through changes in the 'absorption band' diagram (fig. 5.5(a) and 5.10) to the 'energy transfer' diagram [fig. 5.5(b)].

5.7. The observation of metastable decompositions.

It may be observed from the discussion of rates of reaction, and particularly from the equation given in §5.5, that close to the appearance potential of each ion the rate constant for the decomposition reaction may have quite low values, for example, 10^5 s^{-1}. This implies a mean lifetime for such ions of about 10^{-5} s, which is comparable with the time needed for analysis in a typical analytical mass spectrometer (§**3.2**). (In fact the equation is incorrect in suggesting that values of k may be made arbitrarily low by reducing electron energy. If the quantum conditions are correctly applied,

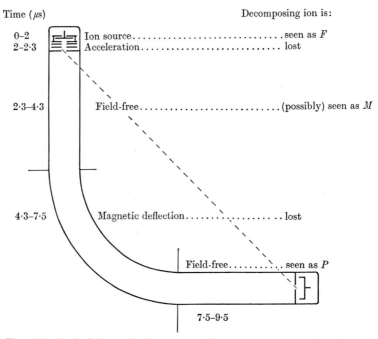

Fig. 5.11. Typical time scale in a sector field mass spectrometer (for 150 cm radius, mass 44 ion, accelerated to 1240 eV).

it is seen that a certain minimum rate constant exists, for excitation to the ground state of the complex, which should often be of the order of 10^5 s^{-1} (Vestal). In such cases, the so called 'metastable peaks' are often observed in the mass spectrum, having a characteristic broadness and apparently non-integral 'mass number'. The intensity is quite dependent on instrumental factors.

In setting out an explanation of the effect, it is worth repeating the remark of at least one other author, that the ions observed at the 'metastable peak' are no longer metastable, but are the products

5.7. Observation of metastable decompositions

of such a decomposition, which has occurred during the flight of the ion through the instrument. A typical time scale for ions passing through a single-focusing sector instrument is shown in fig. 5.11.

The point to remember is that, when decomposition occurs, the instantaneous velocity of the ion is (nearly) unchanged but its momentum and its acceleration, if any, do change. For this reason, the products arrive at no well-defined part of the mass spectrum if decomposition occurs during the main acceleration ($t = 2\cdot0$–$2\cdot3\,\mu\mathrm{s}$) or in the magnetic field ($t = 4\cdot3$–$7\cdot5\,\mu\mathrm{s}$), and such ions are effectively lost. If the change occurs after the magnetic analysis, the fragment ion is detected and counted as if it were a parent ion. Only if the parent ion decomposes in the source, where acceleration is very slight, is the fragment ion seen as such. Decomposition in the field-free region (here, $2\cdot3$–$4\cdot3\,\mu\mathrm{s}$) leads to the detection of a 'metastable peak' as will now be shown.

As may be recalled from §3.2, the acceleration of the ion gives it a velocity of 'v' expressed by $\tfrac{1}{2}mv^2 = eV$ while the magnetic deflection is governed by the law $mv^2/r = Bev$. Hence

$$v = \left[\frac{2eV}{m}\right]^{\tfrac{1}{2}} = \frac{Ber}{m}$$

Thus the condition for detection of an ion may be expressed as

$$r = \frac{m}{eB} \times \left[\frac{2eV}{m}\right]^{\tfrac{1}{2}} = \frac{1}{B}\left(\frac{2mV}{e}\right)^{\tfrac{1}{2}}$$

Now if the ion is accelerated as with mass m_1 and analysed with mass m_2, we may write

$$r = \frac{m_2}{eB} \times \left(\frac{2eV}{m_1}\right)^{\tfrac{1}{2}} = \frac{1}{B}\left(\frac{2m^*V}{e}\right)^{\tfrac{1}{2}}$$

Thus it is seen that the 'metastable peak' appears at an apparent mass

$$m^* = \frac{m_2^2}{m_1}$$

This shows that the observed value of m^* will provide unique confirmation of the mechanistic relation between two other peaks, which will be observed in the spectrum at their normal mass numbers. Unfortunately, such a check of the breakdown scheme is not invariably found for each decomposition reaction, so the failure to observe an appropriate m^* cannot be taken as evidence that the proposed step does not occur.

When using a double-focusing mass spectrometer [see §3.3 and fig. 3.3(b)], 'metastable peaks' can be observed for ions which undergo decomposition between the electrostatic and magnetic analysers. This situation is no different in principle from that just

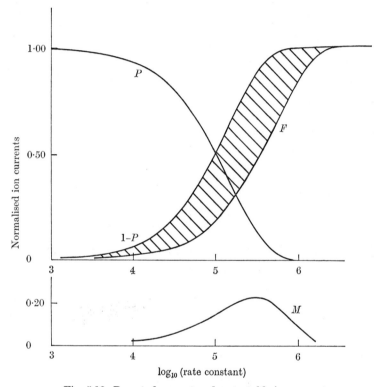

Fig. 5.12. Parent, fragment and metastable ion currents versus $\log_{10}$ (rate constant).

discussed. However with these instruments a second mode of operation exists, and is proving increasingly useful. Ions which decompose in the field-free region immediately preceding the electrostatic analyser will enter the latter with an energy which is lower (by the ratio new mass/former mass) than the energy of the majority of the ions. The energy analyser is readjusted so that it passes the 'metastable' ions and rejects all the ions of 'normal'

5.7. Observation of metastable decompositions

energy. The magnetic analyser then displays a mass spectrum of metastable ions (of chosen energy) free from the overburden of the normal mass spectrum. The quantity of information which may be derived in this manner should go far towards providing firm knowledge of 'breakdown schemes' (Jennings).

The characteristic broadening of 'metastable peaks' in the spectrum is due to small changes of velocity occurring when the parent ion decomposes, and for the same reason the collection efficiency for these ions may be low. Other factors governing the intensity of the metastable peak, which is always small, may be considered as follows. Suppose that unit current of parent ion is generated in the ion source of fig. 5.11 at time $t = 0$. Taking the rate constant of the decomposition as a variable quantity, the resulting ion currents observed at the collector are reliably calculable; these are shown in fig. 5.12. With increase of rate constant the parent ion current 'P' falls as an increasingly large fraction of it is able to decompose within the stipulated $7 \cdot 5 \,\mu$s. The fragment ion current F, is that produced within $2 \,\mu$s. The shaded area between F and $(1 - P)$ represents ions which are lost, except that some are observed as 'metastable', as shown at M. Notice first that the metastable peak is observed within quite a narrow range of rate constant values, and secondly that a rate constant of about $5 \times 10^3 \,\mathrm{s}^{-1}$ is required before the current 'F' reaches even 1 per cent of the total.

To consider these effects in relation to electron energy, it must be recalled that log k is very approximately linearly related to the internal energy scale and that a spread of values of k will be generated, as illustrated by fig. 5.5(b). Of the many values of k, only a narrow range contributes to the 'metastable peak' (which is also inefficiently collected) and hence this is always of low intensity. On the other point mentioned, if the current 'F' must reach 0·1 per cent of the total before an 'appearance' is registered, then about 10 per cent of ions must have reached the rate constant of $5 \times 10^3 \,\mathrm{s}^{-1}$ given above. This may be seen to mean that the observed A.P. will be systematically higher than the true value, an error which is referred to as the 'kinetic shift', and which may amount to 0·1–0·2 eV in many cases. (Fig. 5.5(b) exaggerates this effect.)

The factors discussed above are obviously dependent on instrumental factors, and the broad outline is confirmed by experiments. It should be noticed that the discussion is confined to sector field instruments, and that similar phenomena would not be expected in time-of-flight or radio-frequency instruments, at least in their normal modes of operation.

5.8. A summing-up. The comparisons of experimental data with calculations, while admittedly leaving room for improvement, are generally regarded as providing powerful encouragement to further refinement of the work in this field. The general principles of the model seem sound, and a variety of features, including those of ion source temperature and isotope substitution of the substance studied, can be included in the discussion.

It seems useful to review the assumptions and approximations which have been made, and which must be attacked to provide further improvement of the calculations.

(a) The linear threshold law—this requires confirmation for larger molecules at low electron energies, and further examination for all molecules with electron energies up to the 70 eV region.

(b) 'Vertical ionisation'—the assumption of the Franck–Condon principle seems fairly safe.

(c) Rapid interchange between all states of the same energy—an essential assumption of the theory which, while generally seeming to be true, may prove to have important exceptions. The assumption also makes irrelevant for this purpose any consideration of which electron is removed during the act of ionisation.

(d) The integral approximation—the assumption of a continuum of levels can (and in most cases should) be replaced by a more involved calculation.

(e) The use of parameters of the neutral molecule to describe the ion and its activated states—so far this is the best that can be done.

(f) The omission of rotational effects—they can be included.

(g) Uncertainty of values of E_a—these can only be taken as differences between two appearance potentials or of the appearance and ionisation potential, which are known to rather poor accuracy. One source of error is 'kinetic shift', but thermal excitation of the neutral molecules will give a counter-error. Much better accuracy is needed.

5.8. A summing-up

(h) Ambiguities in and simplifications of the breakdown scheme —more confirmations by 'metastable peaks' may be found; more involved calculations may be carried out.

(i) the absorption band diagram—this is very subject to experimental error, being the sum of several second derivative curves. The description given accounts only for ionisation from the lowest level of the neutral molecule, and extension to cover further vibrational states might be needed, perhaps as a separate absorption band diagram for each vibrational state which is likely to be appreciably populated.

5.9. The empirical approach.

Mass number determination. The empirical approach to the study of mass spectra calls on the large body of experience now accumulated and from a study, firstly of the mass numbers represented in the spectrum, and secondly of their relative abundance, deduces the probable molecular weight and structure of the compound studied.

The nomenclature in this field still seems imperfectly standardised and the terms used here (following several other authors) are as follows.

The *parent ion* is the ion derived directly from the molecule (or rarely, radical) introduced into the source.

A *molecule ion* is the charged form of a recognized stable molecule, or similar species, e.g. N_2^+, CH_4^+, $C_2H_4^+$, H_2O^+.

A *radical ion* is the charged form of an entity regarded chemically as a radical, e.g. NH^+, CH^+ CH_3^+ $C_2H_5^+$, OH^+.

A *fragment ion* (which may also be either of the preceding two types) is one resulting from a simple bond rupture in some heavier ion, while a *rearrangement ion* results, usually from a heavier ion, by more complicated processes which are appropriately rarer.

For covalent organic compounds involving the most common elements (C, H, O) it can be seen to be true that the molecular weight for the most abundant isotopes will be *even*, and the parent ion must show an even mass number, symbolised in fig. 5.13 as (even+). Following this line of thought one may generate the array of possible processes shown in fig. 5.13. As will be discussed, the processes leading to ions of even mass number (types III and IV, besides ionisation) are less common than the others; hence the first search of a mass spectrum is for even mass numbers. It is obviously important to locate the parent ion since this will give the molecular weight of the compound, and location of rearrangement ions (if any) can give important clues to the structure. Almost the only distortion of the sense of this system occurs when the original substance has an odd number of nitrogen atoms, having then an odd molecular weight.

General patterns of decomposition. This short section cannot, and does not purport to give a complete systematic study of this topic, which is fully treated in books devoted entirely to the subject

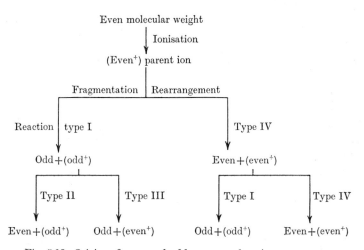

Fig. 5.13. Origins of even and odd mass numbers in mass spectra.

(Fleming & Williams; Budzikiewicz *et al.*). Its purpose is more to indicate the lines along which the observations are rationalised, and deductions made.

(*a*) Fragmentation effects. It is not surprising to find that bonds which appear to break most readily to give fragment ions are those bonds which are weak and 'reactive' in chemical terms, and *vice versa*. Thus in the simplest hydrocarbons, the n-paraffins, all C—C bonds are very similar, and the fragmentation is not strongly biassed in favour of any one mode (though the probabilities are *not* equal). However, in the case of branched paraffins there is a marked preference for fragmentation at the branches. The double bond of

5.9. The empirical approach

olefins has a strong directive effect, the C—C bond next-but-one to it (indicated as 'β') being much more likely to break than an α-bond. The opposite trend is seen in molecules containing the aromatic nucleus, where the chemical stability of this is similarly reflected in the relatively rare fragmentation of this portion of a molecular ion, and generally an increased stability of all ions embodying it, notably the parent ions. Similar stability is noted in parent ions of cyclic paraffins.

The introduction of heteroatoms (most often O, but also N, S, halogens, etc.) into the molecule affects the mass spectrum markedly. Besides giving the possibility of ion masses not seen from hydrocarbons, there is now a strong 'focus' for ionic charge, so that it is often considered to reside entirely on the heteroatom.

The most prominent fragmentation seen in compounds with an oxygen atom is generally the 'α-cleavage', so that for aldehydes or ketones the ions seen are:

aldehydes: $\underset{H}{\overset{R}{>}}C=O^+ \longrightarrow HC=O^+$ (preferred in low molecular weight aldehydes) or $RC=O^+$

ketones: $\underset{R'}{\overset{R}{>}}C=O^+ \longrightarrow \begin{matrix} RC=O^+ \\ R'C=O^+ \end{matrix}$ the lower molecular weight is preferred

Similar reactions can be noted for esters, alcohols (most notably tertiary alcohols) and ethers. In the case of amines and amides similar reactions can be traced, assuming the charge to be carried by the nitrogen atom.

The rough parallels between chemical activity and mode of fragmentation shown above can be understood qualitatively in terms of the relative bond strengths in the neutral molecule or reacting ion. Another important influence on fragmentation is the stability of the ion which can be formed, as is strikingly shown in the following examples. The ethylene ketals can be formed from ketones and show a very strongly controlled cleavage due to the 'resonance stability' of the resulting ion.

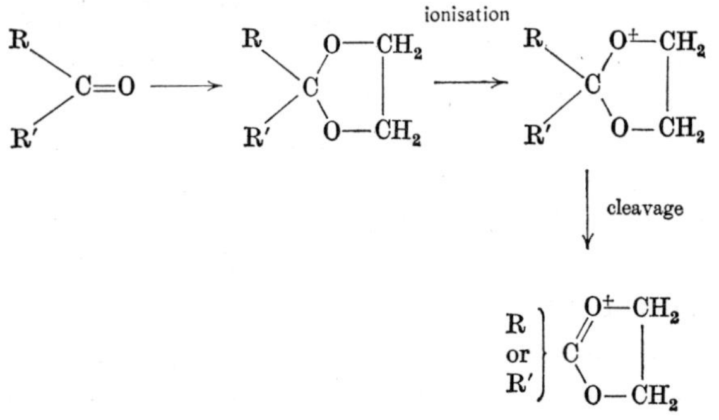

[Diagram: ketal formation from R₂C=O showing ionisation to form cyclic dioxolane radical cation, followed by cleavage to give either R or R' loss with retention of the O±—CH₂ / O—CH₂ ring]

The fragmentation of a long-chain alkyl halide often gives a strong contribution of $C_4H_8X^+$ which probably has a ring structure

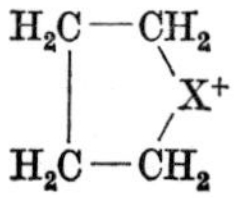

[Structure: cyclic $\begin{array}{c} H_2C-CH_2 \\ H_2C-CH_2 \end{array} X^+$]

The ionisation of aromatic alkyls gives a prominent fragment of mass 91, which is shown not to be benzyl ion $C_6H_5\text{—}CH_2^+$, but the cyclic tropylium ion

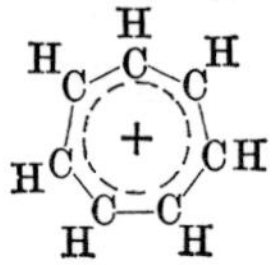

[Structure: tropylium cation, seven-membered ring with + charge]

(b) *Rearrangement.* One of the most commonly-found rearrangements is that involving 'β-cleavage' in which a hydrogen atom is apparently transferred during the process. The mechanism probably involves an intermediate six-membered ring, for example

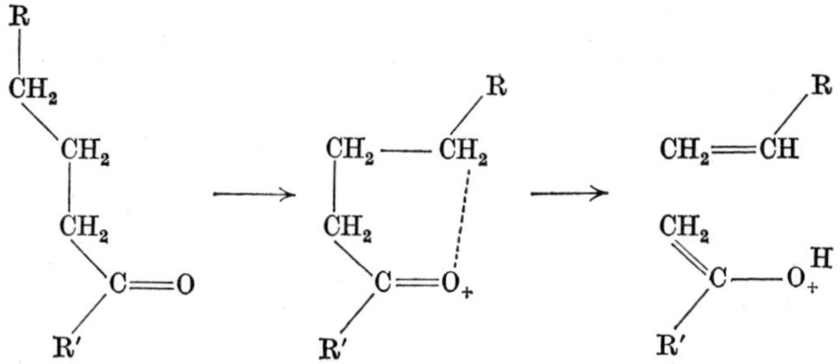

[Diagram: McLafferty rearrangement showing R—CH₂—CH₂—CH₂—C(=O)R' → six-membered transition state → CH₂=CH(R) + CH₂=C(OH)R'⁺]

Thus ketones, aldehydes, esters and amides of sufficient chain length lose an olefin fragment and give an ion of even mass

5.9. The empirical approach

number. Acetates appear to show unusual behaviour in that the same rearrangement eliminates acetic acid and leaves the charge on the olefin fragment. However, the loss of water from long-chain alcohol ions is very similar in character, and one may note that acetic acid and water are very stable entities. The aromatic nucleus provides a good basis for such intermediate cyclic states, and examples may be cited for long-chain alkyl benzenes.

and for *o*-substituted aromatic ethers

Some instances are found in which a rearrangement follows a fragmentation process as in ethers

and in some amines

These last two examples might be regarded as indicating that the intermediate may have a ring of other than six atoms. An alternative and perhaps preferable view is simply that of a hydrogen atom transfer with no commitment to the idea of a 'ring'.

(c) *Other aids in diagnosis.* It is already recognized that the occurrence of ion–molecule reactions, as a result of judicious elevation of source pressure, can sometimes be of assistance in the analysis. In the case of amines, the peak at $18(NH_4^+)$ is a good diagnostic feature, and likewise at $19(H_3O^+)$ in the case of alcohols, though this is more susceptible to interference from water traces. However, a notable feature of alochols is the difficulty of finding a parent ion peak. Elevation of the pressure usually yields, by ion–molecule reaction, a recognisable $(P+1)$ peak, from which the molecular weight is then known.

The truth of many of the postulated modes of fragmentation may be afforded direct evidence by the observation of appropriate metastable peaks, which are more and more being employed in this way. Sometimes the observation gives evidence of a mass peak (especially a parent ion peak) which is hardly observable in the spectrum.

5.10. A reconciliation. The building of a bridge between the theoretical treatment of mass spectra and their practical use in qualitative analysis is surely the province of the physical chemist. He can hope to offer experimental data to the former and to crystallise and clarify the rationalisation of the empirical approach. This section attempts to show the foundations on which such a bridge might be built.

The ionisation process. The first remark which may be made here is that the electron cloud of a molecule is sufficiently mobile that the readjustments of it due to removal of one electron will take place very rapidly, even during the ionisation process. The electron is thus rightly considered to be removed from a molecular orbital, which is presumed to be the one highest in energy. More extensive and more accurate data on ionisation potentials are needed in this approach, which has already shown for long-chain molecules an encouraging correlation between calculated electron densities in some half-filled orbitals and relative frequency of bond rupture in an ion.

Another important factor is the general decrease of stability of a parent ion, compared with the molecule. This arises since the

5.10. A reconciliation

electron is generally removed from a bonding orbital, leaving an unpaired electron—parent ions are generally odd-electron species (though as noted previously, generally of even mass number). The whole question of bond strengths in ions, or of the vibration frequencies, is largely unexplored, and valuable contributions may be expected if spectroscopic techniques can be extended to this domain.

Lastly one may note that in the presence of 'hetero-atoms', an electron will be most easily removed from the non-bonding orbitals closely associated with them, giving reasonable ground for taking them as the centre of charge. However, the possibility of some degree of delocalisation of the electron deficiency must be remembered, especially when this may be influenced by the rest of the molecule through conjugation.

The choice of decomposition path. Three general observations may be made on this topic. Firstly, as the molecule becomes ever larger, the number of possible routes for decomposition will soar, and this fact alone may account for the observed decrease of parent ion peak as higher molecular weights of any one type of compound are examined. Secondly, it is generally true that, in terms of the scheme of fig. 5.11, the (odd$^+$) species are the more abundant. This arises by two assisting influences. The (odd$^+$) ions are even-electron species, and have an inherent advantage of stability in that respect. Also, they are mostly formed via fragmentation reactions rather than the rarer, or slower, rearrangement reactions. A reason for this last observation may in turn be suggested, following the theory of mass spectra previously discussed. The fragmentation reaction proceeds via a transition state which is 'looser' than the original ion (the vibration frequencies are lower), and this may be seen to lead to a high frequency factor, or a positive 'entropy of activation'. The reverse is true for a rearrangement, when the transition state is often pictured as having a cyclic form with at least one more bond than the reactant ion; this may be expressed as a negative 'entropy of activation', acting of course to reduce the rate of reaction by this path. The characteristics of such transition states are still largely a matter of speculation, with suggestions being put forward that electrons and protons can move very freely in them and that the idea of directed valencies may apply only loosely.

Lastly, though often given first place, is the influence of stability of the products on the course of the reaction. Examples have been given in §5.10 where exceptional stability of either the ion or the neutral species has clearly played some part. However, the absolute reaction rate theory does not allow an immediate postulation of a direct connection between rate of a reaction and the change of internal energy (ΔE); obviously the activation energy is the key factor. Nevertheless, since all decomposition reactions involve breaking at least one bond, it is the case that these reactions are almost all endothermic, and the activation energy could be equal to the increase of internal energy. It must be remarked that values of this latter quantity are often sadly lacking in accuracy, if they are known at all.

The question of an energy barrier. The theory of mass spectra shows that rates of reaction may be expected to be extremely sensitive to the actual activation energy of the reaction, which may be equal to the endothermicity *plus* an additional 'energy barrier'. At present this question has only been attacked by comparing heats of formation of ions arising by several different routes. In some cases, values appear to be inconsistent by amounts greater than the expected error, and it is often credible to interpret this as implying an 'energy barrier' in some reactions. The information obtained from such considerations is illustrated in fig. 5.14 and is in fact part of the data used in the calculations of the propane mass spectrum (see §5.6). The figures for heats of formation, appearance potentials etc. which are used still leave a great deal to be desired in accuracy, but it appears to be true that simple bond fractures involve little or no 'energy barrier', over and above endothermicity, while rearrangement reactions show definite energy barriers of up to 10 kcal/mole (rarely, rather more). The latter is not surprising, except perhaps that the values are so small; activation energies of comparable reactions of neutral molecules are much larger.

One definite principle which seems to have emerged is known as Stevenson's Rule (Stevenson). This refers to the fragmentation of a molecule R_1—R_2, assuming that the ionisation potential of R_1 is larger than that of R_2. It is then usually the case that decomposition of the parent ion to $R_1 + R_2^+$ is more probable, and that if the reaction producing R_1^+ occurs, the products in some way carry excess

5.10. A reconciliation

energy. This may be illustrated on an energy diagram as in fig. 5.15, which indicates that the latter reaction has a marked 'energy barrier', while the former has little or none. Thus only reactions of type leading to $R_1 + R_2^+$ may be used in bond-energy calculations (§ **4.4**), p. 85).

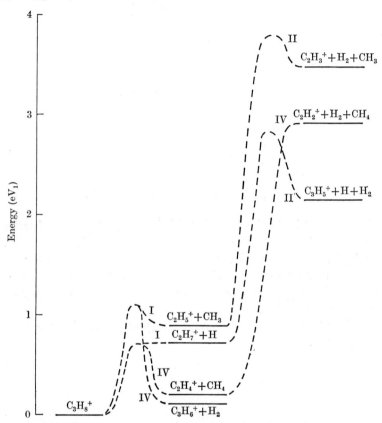

Fig. 5.14. Energy barriers in the fragmentation and rearrangement of propane ions.

It seems that further progress in this aspect must await more accurate measurements of such quantities as appearance and ionisation potentials. It may be appropriate to close with an indication of a possible new line of attack, which is most easily illustrated by a reaction drawn from fig. 5.14 and labelled type IV

$$C_3H_8^+ \rightarrow C_2H_4^+ + CH_4$$

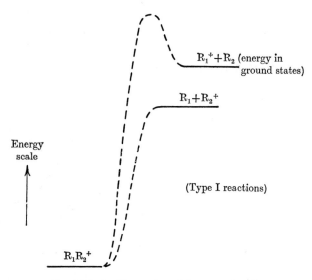

Fig. 5.15. An illustration of Stevenson's rule.

The present indications are that this has a definite energy barrier of about 25 kJ/mole and a small endothermicity of 20·5 kJ/mole. It might be possible to determine the height of the energy barrier more directly by a study of the reverse reaction, under conditions of thermal energy at two or more temperatures—this is simply an ion-molecule reaction.

And that, the conscientious reader may observe, takes one back to chapter 1.

Note on Système International (SI) Units

During the preparation of this book, recommendations have emerged for the adoption of a definite framework of metric units, these being referred to as SI units.

Following this system, thermochemical energy units have been expressed throughout the book as kilojoules per mole (kJ/mole) rather than kilocalories per mole. The more extensive changes which would be necessary for the conversion to SI units of all references to centimetres and Angstrom units of length, and the torr unit of pressure, have not been made. There may moreover be some continuing debate over the general desirability of these particular conversions. Numerical factors relevant to the proposed changes are appended.

SI unit	equivalent in non–SI unit
joule (J)	1/4·184 calorie
	$(1/1·602) \times 10^{19}$ electron volts
metre (m)	10^{10} Angstrom units
	10^2 centimetres
newton/sq.metre (Nm^{-2})	1/133·3 torr

Bibliography and Comprehensive References

A.M.P.I.C. (1969). *Review of Ion–molecule Reactions* (to be published). John Wiley & Sons, N.Y.

Beynon, J. H. (1960). *Mass Spectrometry and Its Applications to Organic Chemistry.* (Elsevier.)

Budzikiewicz, H. Djerassi, C. & Williams, D. H. (1967). *Mass Spectrometry of Organic Compounds.* (Holden–Day).

Duckworth, H. E. (1958). *Mass Spectroscopy* (Cambridge University Press.)

Field, F. H. & Franklin, J. L. (1957). *Electron Impact Phenomena.* (Academic Press), Inc.).

Fleming, I. & Williams, D. H. (1966). *Spectroscopic Methods in Organic Chemistry.* (McGraw Hill.)

Kiser, R. W. (1965). *Introduction to Mass-spectrometry and Its Applications.* (Prentice-Hall Inc.).

McDaniel, E. W. (1964). *Collision Processes in Ionised Gases.* (John Wiley and Sons, Inc.)

McDowell, C. A. (ed.) (1963). *Mass-spectrometry* (McGraw Hill).

McLafferty, F. W. (ed.) (1963). *Mass Spectrometry of Organic Ions.* (Academic Press, Inc.).

Reed, R. I. (ed.) (1965). *Mass Spectrometry* (Academic Press.)

Specific References

Barber, M., Cuthbert, J., Farren, J. & Linnett, J. W. (1963). *Proc. Roy. Soc.* A **274**, 285.
Beckey, H. D., Heising, H. Hey, H. & Metzinger, H. G. (1968). *Adv. Mass Spectrometry*, **4** (Inst. Pet.) (and neighbouring papers).*
Berkowitz, J. (1962). *J. Chem. Phys.* **36**, 2533.
Berkowitz, J. & Marquert, J. R. (1963). *J. Chem. Phys.* **39**, 275.
Berkowitz, J. & Chupka, W. A. (1964). *J. Chem. Phys.* **40**, 287.
Böhme, D. K., Hasted, J. B. & Ong, P. P. (1967). *Chem. Phys. Letters* **1**, 259.
Brehm, B. & Puttkamer von E. in *Adv. Mass Spectrometry*, **4**, (Inst. Pet.).*
Castaing, R. & Slodzian, G. (1962). *J. Microscopie* **1**, 395.
Chan, C. C. & Sedgwick, R. D. (University of Manchester) (to be published).
Chupka, W. A. & Inghram, M. G. (1955). *J. Phys. Chem.* **59** 100.
Clyne, M. A. A. & Thrush, B. A. (1962). *Disc. Faraday Soc.* **33**, 275.
Crasty, R. L. & Mitchell, J. G. (1966). *Earth and Planetary Science Letters* **1**, 121.
Dugan, J. V. & Magee, J. L. (1967) *J. Chem. Phys.* **47**, 3103.
Epstein, S., Buchsbaum, R., Lowenstam, H. A., & Urey, H. C. (1953). *Bull. Geol. Soc. Am.* 1315.
Fehsenfeld, F. C., Schmeltekopf, A. L. & Ferguson, E. E. (1965). *Planetary Space Sci.* **13**, 219.
Gay, I. D., Glass, G. B., Kistiakowsky, G. B. & Niki, H. (1965). *J. Chem. Phys.* **42**, 608; **43**, 4017.
Gomer, R. (1961). *Field Emission and Field Ionisation* p. 64, Harvard University Press.
Hayhurst, A. N. & Sugden, T. M. (1966). *Proc. Roy. Soc.* A **293**, 36.
Homer, *Odyssey* 12 (*ca.* 1000 B.C.).
Hush, N. S. & Pople, J. A. (1963). *Trans. Faraday Soc.* **51**, 600.
Jennings, K. R. (1963). *J. Chem. Phys.* **43**, 4176.
Kaneko, Y. Megill, L. R. & Hasted, J. B. (1966). *J. Chem. Phys.* **45**, 3741.
Kantrowitz, A. & Grey, J. (1951). *Rev. Sci. Instrum.* **22**, 328.
Kemball, C. (1951). *Proc. Roy. Soc.* A **207**, 539.
Kemball, C. (1952). *Proc. Roy. Soc.* A **214**, 413.

Knox, B. E. in *Adv. Mass Spectrometry*, **4** (Inst. Pet.).*
Krauss, M. in *Adv. Mass Spectrometry*, **4** (Inst. Pet.).*
Liebl, H. J. (1967). *J. Appl. Phys.* **13**, 5277.
Light, J. C. (1964). *J. chem. Phys.* **41**, 586.
Light, J. C. (1965). *J. chem. Phys.* **43**, 3209.
Light, J. C. in *Adv. Mass Spectrometry*, **4** (Inst. Pet.).*
Lineweaver, J. L. (1963). *J. appl. Phys.* **34**, 1786.
Lorquet, A. J. & Hamill, W. H. (1963). *J. Phys. Chem.* **67**, 1709.
Maier, W. B. (1965). *J. chem. Phys.* **42**, 1790.
Mann, K. H. & Tickner, A. W. (1960). *J. Phys. Chem.* **64**, 251.
Moran, T. F. & Hamill, W. H. (1963). *J. chem. Phys.* **39**, 1413.
Munson, M. S. B. & Field, F. H. (1966). *J. Am. Chem. Soc.* **88**, 2621.
Narcisi, R. S. & Bailey, A. D. (1965). *J. geophys. Res.* **70**, 3687.
Perrerson, E. & Lindholm, E. (1963). *Ark. for Fys.* **24**, 49.
Srinivasan, R. (1961). *J. Am. chem. Soc.* **83**, 4344.
Stebbings, R. F., Rutherford, J. A. & Turner, B. R. (1966). *J. Geophys. Res.* **71**, 771.
Stevenson, D. P. (1951). *Disc. Faraday Soc.* **10**, 35.
Tal'rose, V. L. & Frankevitch, E. L. (1956). *Doklady Akad. Nauk, SSSR*, **111**, 376.
Urey, H. C. (1947). *J. chem. Soc.* **562**.
Vestal, M. L. (1965). *J. Chem. Phys.* **43**, 1356.
Webster, R. K. (1959). *Adv. Mass Spectrometry*, **1**, 103 (J. D. Waldron, ed.: Pergamon).
Wilmenius, P. & Lindholm, E. (1962). *Ark. for Fys.* **21**, 97.

* This volume now available (1968). Overseas agents are Elsevier. Editor E. Kendrick.

Index

absorption band diagram, 103, 121
absorption coefficient, 4
afterglow studies, 39, 50, 95
analysis
  isotope ratio, 80
  'on-line', 89
  residual gas, 76
  stable samples, 78
  tracer dilution, 80
  unstable samples, 91
angular distribution, 52
appearance potential, 23, 83, 94
  kinetic shift of, 119
attachment (clustering) reactions, 50, 58
atomic mass, accurate measurement, 76
atomic unit of distance, 5

beam studies of reactions, 52
  double beam, 52, 54
bond energies, *see* dissociation energies
breakdown curves, 112
breakdown scheme, 114

centrifugal potential, 9
charge transfer reactions, 40, 53, 55, 112
collision complex, 54
collision diameter, 4, 18
cracking pattern, 49, 56, 79, 98, 112
cross section
  definition, 3, 18
  electron impact, 5, 22
  experimental, 4, 45
  for 'close' collisions, 11, 13, 16
  measurement, 4, 6
  microscopic, 4, 23, 53
  photoionisation, 5, 27

detection of ions
  Faraday cup, 72
  multiplier, 73
  photographic, 31, 63, 74

dissociation energies
  determination of, 83, 97, 129
drift methods for studying reactions, 49

electric discharges
  ions from, 34, 36, 50
  radicals from, 94
  spark source, 31
electron affinity, 86
electron impact ionisation
  cross section, 5
  efficiency curves, 99
  linear threshold in, 99
electron impact ion source, 21, 45, 79
  effect of temperature, 115
  high pressure operation of, 48
  pulsed operation of, 48, 50, 66
endothermic reactions, 3, 8, 16
energy absorption band diagram, 103, 104, 121
energy analyser (selector), 25, 63, 118
  see also velocity
energy barrier, 31, 105, 128
energy levels, density of, 110
energy of activation, 3, 129, 128
energy transfer, 14
  diagram, 103
entropy of activation, 127

field ionisation, 28
flames, ions from, 34, 39
  radicals from, 94
focused radiation ion source, 32
fragmentation of ions, 23, 27, 31, 49, 122 ff.
  α-cleavage, 123, 125
fragment ion, 116, 121

gas-liquid chromatography, 41, 90

heats of dissociation, *see* dissociation energies
heats of formation
  of carbon vapour, 96

heats of formation (cont.)
  of ions, 86
  of oxides, 97
  of radicals, 97

impact parameter, 8, 15
  critical, 10, 16, 18
integral approximation in theory of mass spectra, 111
ionic equilibrium, 40, 59
ionisation efficiency curves, 99
  second derivative of, 100, 103, 112
ionisation potential, 23, 32, 83, 126
  calculations of, 83, 92
  determination of, 25, 27, 105
  of radical species, 94
ion-molecule reactions
  examples, 1, 17, 38, 40, 44, 50, 52, 55 ff.
  mechanism, 54
  tests for, 46
  use in analysis, 126
isotope distribution, 79
isotope ratio determination, 32, 80
  age of rocks by, 81
  temperature of sea by, 82

Knudsen cell, 25, 95

Langevin 'slow ion' theory, 8 ff, 13
leak detection, 75
linear threshold law, 99

mass doublets, 77
mass spectrometer
  cycloidal, 65
  double-focusing, 31, 36, 62, 118
  linear accelerator, 67
  omegatron, 68
  quadrupole, 40, 69
  single-focusing, 60, 116
  time-of-flight, 66
mass spectrum, calculation of, 114
metastable ions, 115 ff, 126
metastable orbit, 11, 12
microprobe ion source, 32
molecular beam, 91
molecular flow of gas, 22, 35, 36, 90
molecular weight
  accurate measurement of, 77
  determination of, 122, 126
molecule ion, 121

parent ion, 23, 121
  breakdown of, 114, 116
  internal energy in, 103
  stability of, 123, 126
phase space, 15
photoelectron spectroscopy, 105
photon impact (photoionisation)
  ion source, 26, 43, 50
  cross section, 27, 100
polarisibility, 8, 14
polarisation forces, 8
proton affinity, 88
proton transfer reactions, 49, 57, 88, 112

quasi-equilibrium theory of mass spectra, 108 ff

radiation chemistry (radiolysis), 1, 56
  α-particle, 48
radical ion, 121
radical species, detection of, 93
rate of decomposition, 108, 111, 118
reaction rate constant
  typical values, 2
  'thermal', 4, 20, 45
  'chemical', 19
rearrangement ion, 121
rearrangement processes, 124 ff
residual gas analysis, 76
resolving power, 60
retarding potential difference (RPD) method, 25

second derivative curves, 100, 103, 112
shock waves, ions from, 34, 41
skimmer, 49, 91
spark source of ions, 31, 81
Stevenson's rule, 128
surface ionisation, 32

temperature effects
  on ion-molecule reaction, 20
  on ion fragmentation, 115
time of collison, 12, 14
tracer dilution, 80
transition time, 14

upper atmosphere, 34, 42
  special mass analysis method, 71

velocity selection (analysis), 52
viscous flow of gas, 35, 90